ANNEXE DE LA BIBLIOTHÈQUE
uOttawa
LIBRARY ANNEX

MAKING TIME

Ethnographies of High-Technology Organizations

MAKING TIME

Ethnographies of High-Technology Organizations

Edited by
Frank A. Dubinskas

Temple University Press
Philadelphia

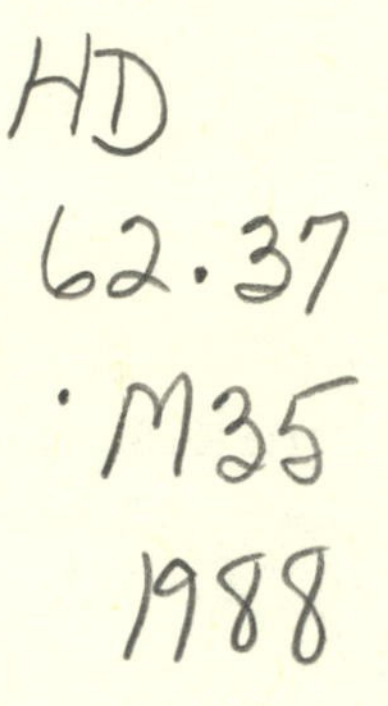

Temple University Press, Philadelphia 19122

Published 1988
Printed in the United States of America

The paper used in this publication meets the minimum requirements of American National Standard for Information Sciences—Permanence of Paper for Printed Library Materials, ANSI Z39.48-1984

LIBRARY OF CONGRESS
Library of Congress Cataloging-in-Publication Data

Making time : ethnographies of high-technology organizations / edited by Frank A. Dubinskas.
p. cm.
Bibliography: p. 229
Includes index.
ISBN 0-87722-535-4 (alk. paper) :
1. High technology industries—Management. I. Dubinskas, Frank A. (Frank Anthony)
HD62.37.M35 1988
620'.0068—dc19 87-28411
CIP

Contents

MAKING TIME

Ethnographies of High-Technology Organizations

CHAPTER 1

Cultural Constructions: The Many Faces of Time

Frank A. Dubinskas

THE DIVERSITY of *times* in high-technology and scientific cultures is the central theme of this collection. Our studies examine the socially constructed character of *time* within different culture-worlds in advanced industrial society. Time—or, better, times—means different things to each of the communities of scientists, engineers, doctors, and executives that we investigate. In speaking of times, we are never talking about a single universal entity, concept, or system; and one of the fundamental conclusions from our studies is that no one group or culture has a monopoly on the definition of time. Yet all these times have a "family resemblance" as important symbolic nexes around which coalesce issues of order, power, self-definition, and knowledge.

These ethnographic studies[1] examine social constructions of time within North American professional cultures. We are accustomed to speaking of them all as parts of "Western Culture," as if there were some uniform context or hegemonic framework that patterns time. It is a central theme of these essays, however, that *differences* in the social construction of times are crucial factors in doing scientific and technical work and in shaping the communities of professionals who do it.

Our subjects are the prestigious and sometimes powerful technical professionals and managers who produce state-of-the-art knowledge as well as sophisticated technical devices and machines. They include physicists investigating the fundamental constituents of matter (Chapter 2), biologists altering the molecular building blocks of genetic material (Chapter 5), semiconductor engineers turning sunlight into electrical power (Chapter 3), and medical professionals creating images of living tissue (Chapter 4). There are several modes of variation within and among these. One is the multiple varieties of time within the same community, like the six kinds of "social time" for particle physicists that Sharon Traweek describes in Chapter 2. Another is the contrasting times of different communities or groups working in the same organization or around the same technology. In these organizations, time is often an articulation point, mediator, or bone of contention between the professional groups (Chapters 3–5).[2]

Our point is that each of the sciences and technologies, embodied in a community of expert practitioners, presents its own particular visage to itself and to the wider worlds of its environment. And within each community, the patterning of time is a central aspect of social order and process, as well as a focal point of meaning and knowledge production. This view stands as a counterpoint to the chorus of rhetoric that collapses all times, all sciences, and all technologies into a uniform "Western Science and Technology"; and they are an anthropological invitation to explore the differences among the ranks of our nearer natives, friends, and colleagues.

TIME AND DIFFERENCE

Anthropologists have a long history of examining cultural diversity in exotic locales, and time has also figured centrally in many ethnographic studies. Our European and North American neighbors are more sparsely represented. For example, the anthropologist Edward T. Hall's synthetic work *The Dance of Life: The Other Dimension of Time* (1984) on "different kinds of time" draws broadly on ethnographic studies by Paul Bohannon (1953), E. E. Evans-Pritchard (1940), and Barbara Tedlock (1981) among the Tiv, Nuer, and Quiche Maya. Hall also discusses his own as well as other research contrasting "Western" and "Eastern" (Oriental) temporal cultures as well as two kinds of European (southern vs. northern) temporal patterns. Hall's concepts of "polychronic" and "monochronic" time treat the issues of multiple and overlapping temporal realities in a fashion reminiscent of Barley's work in this volume. Hall's general principle, that "time is not just an immutable constant, as Newton supposed, but a cluster of concepts, events, and rhythms covering an extremely wide range of phenomena" (1984, p. 13) accords well with our perspectives. He is, however, relatively agnostic about the cross-comparability of his "different streams of time." Also, while Hall introduces the description of "tempo" to his contrasts between northern and southern European time-senses, Pierre Bourdieu (1977), whose work I discuss later, uses tempo analytically in conceiving an important theory of timing and strategic action.

Our anthropological analyses of diversity within the realms of Western sciences and advanced technologies are facilitated by path-breaking intellectual achievements in other fields. Thomas S. Kuhn's *The Structure of Scientific Revolutions* (1962), in particular, opens the door to critical analyses of uniformist or universal "theories" in science, including theories of time. By arguing that sciences experience a series of discontinuous (revolutionary) historical changes in their models of the world, and that such sequential changes are an integral aspect of science as a socially constructed activity, Kuhn presages our arguments for contemporaneous diversity. While Kuhn argued primarily for differences as a sequence of changes, he also notes that there are periods preceding (revolutionary) change when competing theories exist simultaneously. In Kuhn, social processes determine the eventual replacement of one intellectual hegemony by another; and each hegemonic theory is thus relevant only to its own historical context. In the essays in this book, competing socially constructed and validated models of time exist simultaneously, each valid in its own social and cultural context.

Social theorists who deal with time in its classic historical sense are legion; those who treat the history of time itself are far fewer. A classical discussion is E. P. Thompson's "Time, Work-Discipline, and Industrial Capitalism" (1967), where Thompson describes changes in the way time was apprehended before and through the British industrial revolution. An important change accompanies the transformation of time from something that *passes* into something that is *spent*. This "accounting" for time is a relatively new cultural pattern in the seventeenth century, and Thompson links its spread to the precise timing of wage labor. Its ascendance as a dominant model is linked to the development of the factory system in the nineteenth century, public displays of two-handed clocks, and the wide diffusion of pocket timepieces.[3] David Landes (1983) also presents a finely detailed history of the relationship between the development of clocks and the new nineteenth-century social construction of industrial time. Both Landes and Thompson are concerned with an emerging mechanized orderliness of time, and with its construction as a linear, progressive, and regular dimension that comes to permeate modern life.

As a pervasive and dominant metaphor in our everyday folk imagery, this Newtonian linear "arrow of time" also comes to dominate modern scientific discourse. Stephen Jay Gould discusses this mutual infusion of folk and scientific models of time as he follows the history of geological argument and its rhetoric from the seventeenth to the nineteenth centuries. In *Time's Arrow; Time's Cycle: Myth and*

Metaphor in the Discovery of Geological Time (1987), Gould traces the dialectical interplay between these two potent images of time, the arrow and the cycle, through the developing scientific argument over the character of geological time. While he finds both images deeply rooted in the Judeo-Christian tradition, he argues that "the arrow" has gained ascendancy in Western culture, particularly with the rise of modern sciences. That linear "scientistic" metaphor now dominates our everyday folk view of time as well.

Many social scientists have rather uncritically incorporated this orderly time of Newton into their own models, treating time as a background or hidden dimension. It becomes an attribute of the natural universe that is simply there (or "ticking away") as a parameter, marker, or line against which events and activities unfold in an orderly fashion and are then measured by the analyst. Time's accountants are legion not only in the factory but also in the academy. Anthony Giddens, for instance, in *The Constitution of Society* (1984) discusses time and order extensively, drawing on the models of social geographers (*vide* chap. 3, "Time, Space and Regionalization," pp. 110–161). He begins by saying, "Most social analysts treat time and space as mere environments of action and accept unthinkingly the conception of time, as mensurable clock time, characteristic of modern Western culture. . . . Social scientists have failed to construct their thinking around the modes in which social systems are constituted across time-space" (p. 110). However, the dimensional foundations of his models of time are, at root, linear and parametric. They are "time-spaces" in a traditional four-dimensional structural model: space in three dimensions plus time as regular flow or change. His contribution lies in understanding that there are multiple overlapping contexts of relevance for differently constituted temporal orders. The underlying—and unexamined—assumptions of his approach are that these times are commensurate. As with other theorists, there is a linear dimension or progressive movement behind all social life; and, in the end, it is assumed that we really can peg all events conveniently to that same line. Researchers then measure events "against" time, theorists discuss events "in" time, and managers and politicians regulate, manage, and control time—or try to—as a way of controlling activity. Time is reduced to a "timeless" dimensionality, a mere "shadow cast by social action."[4]

This reduction is partly born of the social need to control complexity, as well as activity, particularly with the rise of European and North American industrial capitalism. The great utility of this line of reasoning has been argued for (some would say "demonstrated") by generations of planners from the scions of F. W. Taylor's "scientific management" at the turn of the century to the proponents of "management by objectives" (MBO) in the 1980's.[5] One divides complex tasks and processes into ever smaller parts, measures them according to a universal standard, then reorganizes them into the "most efficient" schedule. What's more, the natural sciences themselves are widely perceived to treat time in this parsimonious, divisible fashion; and many social science studies borrow this fundamental link and its presumed rigor from their kindred scientific disciplines.

If modern industrialization (see, e.g., Thompson, 1967; Landes, 1983) and science impose a hegemonic model of time, and it has diffused into everyday folk discourse, we confront powerful social and cultural forces for the propagation of a unitary temporal order. We shall argue by contrast that this linear, dimensional model of time in the social sciences naturalizes what are fundamentally diverse social and cultural categories.

Rather than beginning with a universal, hegemonic Time, all of the studies in this volume begin with the communities of practitioners—scientists, engineers, doctors, technicians, and managers—for whom time is a salient ordering feature of their social existence. The simple line, the scalar measure as the parameter of passage, will not suffice as a model for any of these communities. Two examples may serve to indicate how we develop alternative, multiple views of time.

Traweek's discussion of the cosmology of time among experimental particle physicists provides an antidote to the uniformist view of temporal lines as a "natural" order. The physicists have a dual temporal cosmology; it includes two fundamental and (in physical theory) irreducible kinds of time. These are "replicable [or relativistic] time, which can be accumulated, and calendrical [or non-relativistic] time, which slowly slips away."

> Time in a non-relativistic setting is simply a marker. It is only a "milestone" marking a sequence of events in space. . . . For example, the experiments conducted by scientists in the laboratory occur in a non-relativistic setting; that is, time functions as a marker. This kinematic description of the world applies when the speed of motion is very small relative to the speed of light.
>
> In a relativistic setting, time becomes another coordinate, interchangeable with space. Time is elevated to a status equal to that of space instead of being simply a milestone helping us to find our way through the environment.

These two times subsume all the other variations of social time that Traweek describes. For American physicists, for instance, beamtime represents recurrent, replicable time. It is a negotiable commodity that can be accumulated and reproduced through access to the accelerator's stream of high energy particles. The other calendrical times—lifetimes of ideas, detectors, and physicists—are ephemeral. Physicists see their own "natural" movement in calendrical time as a process of decay. This image of decay is homologous with the character of the sub-atomic particles being investigated: one of the salient qualities of these particles is their rate of decay. The work of physicists around their detectors then becomes a social process designed to resurrect time, abolish the past, and transform calendrical into replicable time.

The modern physics community reveals an intracultural diversity that cannot be reduced when it makes that critical distinction between relativistic and non-relativistic, or "replicable" versus calendrical time. These times can only be understood as contextually dependent social constructions. Physicists' lives, work, careers, and the tools they build and use—both mechanical and theoretical—are the handles Traweek uses to grasp the special character of times in their professional culture. These experimental particle physicists live in small communities around the huge, costly machines called "particle accelerators." *Up-* and *downtime*, for example, refer to whether the accelerator's beam is currently running, and this "seasonal" difference governs a major contrast in the physicists' daily activities. *Beamtime* is a commodity, that of access to the critical resource of accelerator use, which is the basic nourishment of their careers. Other kinds of time refer to life-cycles and processes of career development.

A second example comes from my study of companies in the biotechnologies industries.[6] Two domains of temporal patterning are significant in each of the contrasting communities of executives and biologists. The temporal domains are *planning* time and *developmental* time. Contrasting styles of presenting and arguing over plans—promises for future work—distinguish executives on the one hand (especially those from the worlds of finance, marketing, and control) from the bench scientists and scientific advisors who have come to the firms from academic molecular biology. Each group's unique cultural style of planning is compared to the ways it is socialized to model its career development process and intellectual progress. Within the professional group, the temporal patterning of work styles and career models shares homologies with the images of knowledge and practice in the wider scientific and business environments. For managers, short-term plans and closed-frame problem solving are linked to a model of the self as a "complete adult" and to a model of a world of finite "economic realities." For scientists, open-ended planning and problem solving are linked to images of continuous personal and career growth as well as a "progressive" image of the infinite expansion of scientific knowledge. In this example, two professional cultures, both of which claim to be paragons of Western rationality, demonstrate radically different and conflicting notions of time.

These two examples, in particular, serve not only to highlight our main theme of the diversity of times but also to challenge the often sub rosa assumption of uniformity in Western culture. In anthropological writing, this is the implicit suggestion that it is primarily "Others" who exhibit difference.[7] The implication of Western uniformity disintegrates immediately, however, upon confrontation with the close encounters of our ethnographic examples.

Among ethnographic studies, our papers employ a broader sociological frame than is common in one other major current in work on science and technology: ethnomethodology and similar micro-sociological analyses. In these studies, each situated social action also constitutes a particular meaning-context; but these contexts are more narrowly focused. Bruno Latour and Steve Woolgar, for instance, describe the social construction of scientific facts in a biological research laboratory (1979). In their argument, scientific facts, such as the existence and character of a particular biological substance, are created in the laboratory out of "specific localised practices" (p. 239)—their "circumstances" or context. Their archetypical example is that discovery of "tryptophan releasing factor," or TRF.[8] Examining that laboratory work intimately, they deconstruct it into a fine-grained filigree of circumstantial relationships. These relationships constitute the process and substance of making scientific facts. The facts are not made by or "in" nature; they are built in the lab. The unpacking of that construction process, for Latour and Woolgar, is focused on the "fact" (the substance, its property, its discovery, etc.). Meaning lies in the context itself; but that context is often a minutely localized perspective: conversations at the lab bench, criticisms of a published paper (and its author's reputation), comments on the emerging paper trace from a laboratory chart recorder.[9]

The ethnomethodologists' concern with the epistemological status of scientific claims leads to a convergent focus on issues of interest *within* the arguments of science. The native concerns of the scientist, like "How do you demonstrate the existence of TRF?," also provide the topic for the social science analysts. The establishment of TRF or the discovery of the optical pulsar (Garfinkel et al., 1981)[10] are good examples. The evidence that ethnomethodologists use to build their own claims, however, is often restricted to a limited set of inferences about conversational interaction. Only that talk that seems focused on an immediate or unfolding scientific claim is noted. Even when the scope of evidence is expanded, as it is in Latour and Woolgar's discussion of professional careers and the circulation of rewards and credibility, it still centers on an analysis of the currency of scientific claims as the medium of social exchange (1979, chap. 5).[11]

By contrast, we speak more broadly of the cultures and practices of our subjects. We argue that the appropriate contexts for understanding times, sciences, and technologies include ones broader than just those issues and claims of fact that the native practitioners ascribe to "science" per se—or to medicine, or engineering design, or industrial biotechnology research. We change the focus from the single topic of scientific claims, looking beyond the local construals of relevance that are native to physical, medical, engineering, biological, or economic theories, in order to locate them in wider contexts of social action that also form them.[12] Our studies employ broad comparison to the configuration of relationships across narrower contexts and singular events.[13]

This expanded scale allows us to address different issues in the practice of technical and managerial professions. Larry Bucciarelli looks at the structure and construction of time in the design of solar energy cells as a more general model for the engineering design process. Sharon Traweek examines the building of physicists' careers for their patterned homology with the lives of scientific ideas, research groups, and machines. Steve Barley focuses on the central professional activities of hospital technicians and radiologists—making and interpreting visual images of the body—and he elicits the characteristics of temporal order that facilitate and impede joint work. The context is their professional interactions over machines, and their activities vary with the three different imaging technologies they use. What we observe is social action, and what we interpret in that action is patterning—the shaping of time and the temporal shaping of scientific, technical, medical, and managerial work.

TIME AND ORDER

Time is a fundamental symbolic category that we use for talking about the orderliness of social life. In a modern organization, just as in an agrarian society, time appears to impose a structure of work days, calendars, careers, and life-cycles that we learn and live in as part of our cultures. This temporal order has an "already-made" character of naturalness to it, a model of the way things are. Time is also a guide by which social life is actively and intentionally shaped, a model for action. Schedules and plans are created and changed; holidays and deadlines are fixed or erased; histories and reports are inscribed.

Time, though, is also a category of action, a fuel for the energetics of social life. It is spent in work and passed in pleasure, fleetingly quick or ponderously slow; it is stolen from death and reapportioned to all sorts of human ends. Time is full of the practical activities of everyday life; and through doing things, its periods, points, and cycles become meaningful as lived experience. Deadlines for completion become milestones of progress, time is "taken" and used up, and work gets done "in time." From career paths to experimental protocols, to business plans, the orderliness of time becomes orderly as people live and accomplish things. And, finally, they live and accomplish them with a flexible craftwork, the artifice of building the temporal edifice. This is what Pierre Bourdieu (1977) calls "the strategic manipulation of time"—the setting of tempo, the stretching of boundaries, the rushing and relaxing of schedules, and the celebration of passages. This artful manipulation of time is part of the practical and intentional reconstruction of orderliness. The ability or power to exercise this art skillfully, in a recognizably patterned but not rigidly rule-bound way, is a key to the process of building effective social relations.

While all these ways of considering order and time are interrelated, we will tease them apart in what follows to consider one perspective at a time, before recombining them.

First, let us consider that time that appears to impose structure or order upon the world. Anthropological literature is replete with studies of time and culture in more exotic locales. The *locus classicus* in the discipline for beginning these discussions is E. E. Evans-Pritchard's (1940) description of time and its cyclical patterns, based on seasonal floods and their attendant migrations, which order Nuer social life, ecology, and polity.[14] For the nearer natives in the high-technology labs and companies next door, the ordering times are the calendars and business plans, career cycles and research protocols, that structure the lives and daily practices of scientists and engineers, managers and doctors.

As a pervasive aspect of pattern and order, time is most often treated in the social analysis of advanced industrial societies as a structural element. An interest in order (as both structure and process), for instance, is the main thrust of Giddens' theoretical model of time; and he is particularly concerned with routinized, regular action. Eviator Zarubavel's study on "socio-temporal structures" is one of a handful of modern sociological works that is centrally concerned with time. Zarubavel is particularly interested in those systems of temporal order that are highly regularized, as in *Hidden Rhythms* (1981), whose focus is on "socio-temporal patterns" that build "rigidification" into social action (p. 2). He discusses the social construction of "normalcy" through what he calls "temporal anchoring." Through anchoring, time provides a fixed framework that regulates social action, as well as a cosmological framework for understanding this regulation as a (God-)given order of the social universe. This phenomenon of "givenness" is akin to what I earlier called the "naturalizing" of temporal categories. This structuring time in Zarubavel primarily appears as a background or "parameter" against which social action is conducted and measured (p. 20). This foreground-to-background relationship is perhaps partly an artifact of those specific, highly regularized temporal patterns that Zarubavel (1979, 1981) has chosen to investigate; hospitals, Benedictine monasteries, and religious and revolutionary calendars. A similar sense of naturalized orderliness pervades several of the organizational environments treated in our studies.

Investigating the social and technical contexts of temporal rigidity, Barley examines the intimate details of social practice of technicians and radiologists in two modern U.S. hospitals. In the process, he looks at how each group manipulates and orders its time and activities—or tries to—while working with the other. Three different kinds of medical imaging technologies figure in his study. The relations between radiologists (medical doctors) and technicians using x-rays, ultrasound, and CT (computerized tomography) scanners are marked by conflicts rooted in asymmetries between their "temporal subcultures." Their relationships are also colored by a power difference, as they are for managers over scientists in biotechnology firms, and turn around important machines, as they do for the particle physicists. Following Zarubavel (1981), Barley treats the "temporal structure" of each group's work as a social process of imposing order. This order is, in one sense, a linear dimension, pre-patterned by a work-day clock and "mapped by sequences, durations, temporal locations, and rates of recurrence" (p. 2). In the contexts of using x-rays and, to a lesser extent, ultrasound imaging, these temporal structures are radically discordant between the two groups. Within each group, though, its own particular temporal structure provides a shared social scheme or framework for interpreting the events and actions of daily work. At the same time, though, Barley reminds us that these socio-temporal structures are still cultural constructions, no matter how much they may seem to be mapped onto the apparently "natural" order of clock time.

> Although sometimes loosely tied to physical or biological timetables, socio-temporal orders *carry the force of an objective presence* even when they are divorced from nature. In fact, one of the most potent techniques we humans have for turning culturally arbitrary behavior into social fact consists of our tendency to treat even self-imposed temporal boundaries as inviolable external constraints. (emphasis added)

The discord and difference between these temporal structures is also intimately connected with the character of the machine technologies around which radiologists and technicians work, a topic central to Barley's study and to which we will momentarily return.

As the scale and complexity of post-industrial societies and their organizations increase, the means of finding or creating order within them and the difficulty of doing so seems to multiply. Even as computers help to track this complexity, the very presence of these tools invites further growth and intricacy. Contemporary organizations in science, technology, and business have vast needs for coordination, scheduling, and order, which they often claim are unmet. In addition, their engagement with rapidly changing new technologies spins a yet denser web of electronically mediated (or generated) information, communications, and expertise.

This complexity presents a massive practical difficulty when it is embodied in a large-scale, multifunctional research and development or engineering design project. New product development in the automobile industry is probably one of the most complex systems for building civilian consumer products.[15] It involves coordination among corporate planning, marketing, finance, body styling, product design engineering, tool and die design, and manufacturing (process) engineering. Among the product engineers alone, there are as many disparate specialties as there are systems in a car. Furthermore, some technical tasks, performed by specialists, must precede others in a relatively fixed sequence.

In these automobile projects, time is treated first as a system of measurement. There is one master schedule, with checkpoints, approval dates, and milestones from the first "concept approval" to completion of the first saleable vehicle. A normal span is four to six years of work. Within this schedule, there are multiple layers of more specific schedules, one for each group at each level of subdivision. Not only do these schedules chart the work of particular groups, but they also chart their relations with groups conducting parallel activities, and they connect beginnings and endings of groups that work in sequence.

The imposition of order through schedules is a theme that is expanded in different variations among our papers. Bucciarelli examines the multiple nesting of schedules in a design project: schedules for each project, schedules for each person (who may be simultaneously involved in multiple projects), and the scheduling of the overall load of projects across a whole engineering division (or company)—a "mountain load chart." Bucciarelli, however, finds temporal and design flexibility much more evident in the solar energy panel project than we would expect in automobile design. The number of people and components was lower, the range of expertise was much narrower in scope, and the design process took place in a largely face-to-face environment. Schedule building was intimately tied to the progressive emergence of the artifact—the panel—and the project direction was perhaps less clearly defined (or constrained) technically than new car development is. One result was that schedule formation became a regularized forum for negotiations about the order of work and the character of the artifact.

In the auto companies, too, however, this kind of negotiation process takes place at lower levels in the design hierarchy. The problem for car projects is that when local level adjustments add up to a cumulative change in the progress of the whole, a change at the highest level of schedule cannot be tolerated, for important financial and marketing reasons. In the resulting conflict, either design quality or completion dates must often be sacrificed.

Very large and bureaucratic projects thus appear more rigid, at least at the highest levels, than smaller-scale ones:

Barley's research, like Bucciarelli's, deals with small groups of professionals—radiologists and technicians—trying to coordinate their schedules. From the lower-status technologists' point of view, the doctors impose their loosely structured anarchy of alternating tasks onto the more regular schedules of x-ray technicians with little success. The difference is not just the technical character of doctors' work or the pivotal role of machines, which will be explored later, but rather the asymmetry of power between the groups. The inflexibility of scheduling for auto design engineers can also be viewed as an issue of power, as well as organizational size.

The essays in this book place strong emphasis on patterns for shaping time—for ordering the construction process—as well as patterns expressing the orderliness of time. This distinction is made by Clifford Geertz (1973d) in his discussion of key religious symbols and their meanings. Key symbols are both *models of* something (e.g., the character of deity) and, simultaneously, *models for* something (e.g., guides to righteous conduct). The trick is not only to make the distinction, for it is one with many analogous dual Western ancestors—theory and practice, belief and ritual, and so on—but to understand that this "model of/model for" expression is a way of denoting an intimate reciprocity in the character of such meaningful categories. They are at once the shaping and the shape. Time is not prior to social life, but rather it is made out of social action. Time, culture, and the ongoing processes of organizational life in which they are embedded are inseparable.

Traweek and I, for instance, discuss pattern-building social processes as the reciprocal side of cultural patterns that otherwise appear as already shaped regulators of action. Ethnographic research has a special advantage in uncovering this praxis, because the investigators focus on how things are done, as well as discovering that they are done. Discovering the "how" of practice is at the heart of Clifford Geertz's comment, "If you want to understand what a science is, you should look at what the practitioners of it do" (1973, p. 5). We could just as well substitute "time" for "science."

The examination of social processes and practice has a long history in ethnographic research and writing, and the relationship of praxis to cultural pattern and social order is also the subject of a resurgence of interest in ethnographic theory. As ethnography has moved away from static structural-functionalist, structuralist, and cognitive studies, analyses of the relationship of knowledge to action have expanded. A variety of competing schools increasingly treat historical change on a large scale and practical action on a smaller one as critical aspects of description and theory building. Geertz's work on the symbolization of social action and on the symbol as a model *for* conduct has helped stimulate renewed interest in semiotic approaches. In this vein, Geertz's own work elegantly enshrines action in "patterns" that arrest, generalize, and display a kind of stopped-frame image of typical motion—a verbal pose in the ethnographer's camera. An example of this style is his essay "Person, Time, and Conduct in Bali" (1973c), where he elaborates the notions of personhood and time in their Balinese peculiarities. Time, in that work, is an ordering or structuring principle of cyclical reiteration, rather than an active, negotiable, energetic, or fluid one (a distinction whose importance we will come to in a moment). "Conduct," however, is a much drier and fixed characterization of moods, styles, and their contexts than, for example, the bloody excitement of the Balinese cockfights he describes elsewhere.[16] Geertz's image of Balinese "conduct" is meant to illustrate three key aspects of social interaction: formality ("ceremony"), "stage fright," and "the absence of climax." These convey, in the end, a bloodless image washed free of spontaneity, strategy, or passion, but an image, nonetheless, of an ethos structured out of replicated actions. This is an image of patterned practice that seems too inflexible and standardized to account for the tactical inventiveness of human action—the common improvisations to a cultural choreography.

Among contemporary anthropologists who have explored this flexible maleability of time in more detail is Pierre Bourdieu. In his *Outline of a Theory of Practice* (1977), he develops the concept of a "dialectic of strategies" as a linking theory among the intentionality of actors, "given" temporal structures (and social relationships), and the criticality of time and timing in the process ("practice" to Bourdieu) of doing something. His approach to the interactive relationships among structure, practice, and strategy fits the practical complexities of the organizational environments that our studies examine.

Bourdieu (1977) discusses the importance of the practical manipulation of time in two particular contexts of Kabylian (Algerian) social relations: gift giving and vengeance. A crucial social skill in these relationships is manipulating their tempo—the organization of planned temporal spaces between acts and the speed with which acts are accomplished. The timing of a reciprocal gift is a means for creating or altering the social relations of power, debt, affection, or prestige. "Time derives its efficacy from the state of the structure of relations within which it comes to play" (p. 7). "Play" is a critical element in this description, because it implies the calculable flexibility and overt intentionality of people in manipulating time. "Strategic manipulation" is the means by which this apparently dimensional, linear time is turned into a tool that both creates and symbolizes social relationships. A culturally structured context or occasion, be it strictly ritualized or only loosely patterned, provides the format for "playing on the time, or rather the *tempo,* of the action" (p. 7). Such occasions

> offer unlimited scope for strategies exploiting the possibilities offered by manipulation of the tempo of the action—holding back or putting off, maintaining suspense or expectation, or on the other hand, hurrying, hustling, surprising, and stealing a march, not to mention the art of ostentatiously giving time ("devoting one's time to someone") or withholding it ("no time to spare"). We know, for example, how much advantage the holder of a transmissible power [or knowledge] can derive from the art of delaying transmission and keeping others in the dark as to his ultimate intentions. (p. 7)

Bourdieu then compares this model of practical temporal ordering with the reified images of fixed structure that are more common to mechanistic descriptions of social norms, rules, and rituals:

> This takes us a long way from the objectivist model of the mechanical interlocking of preregulated actions. . . . Only a virtuoso with a perfect command of his "art of living" can play on all the resources inherent in the [temporal] ambiguities and uncertainties of behavior and situation in order to produce the actions appropriate to each case. (p. 8)

"Necessary improvisation" is the art of the strategic manipulation of time and its ambiguities in Bourdieu, and this kind of loosely choreographed or stylized improvisation is particularly relevant to the realms of time discussed in our essays. Bucciarelli, for example, discusses the constant changes in plans and their schedules, as the design process proceeds. The differences in working plans (and in images of the emerging panel) between subgroups of the design team facilitate design creativity. A dynamic tension is created by uncertainties about the next best step(s) to take, and the process of ongoing negotiations over them provides a forum for building design convergence toward a "completed" artifact.

Bucciarelli argues that, contrary to the manager's desire to circumscribe, plan, and "chart out" the beginnings and ends of projects, a notion like "beginning" or completion" must remain fairly indistinct. The design team may be finished with the panel when a prototype is finally built and passed on to manufacturing, but the artifact may go through numerous subsequent revisions—not all at the hands of those product designers. Also, not every component task group of the design team "finishes" their work at the same time; they might rather stagger to a conclusion asynchronously. Even if everyone "signs off" formally on the project at the same time, "completion" is an ambiguous quality of the object and its creation process. The end is socially fixed by the sign-off approval. That ending can be tactically manipulated—for example, held up or sped up—depending upon the social aims and power of different members of the group; or the group as a whole may manipulate design closure vis-à-vis another group, like manufacturing engineering.

Small-scale observations provide critical details of the daily practices through which time is built up as a meaningful ordering category.[17] Bucciarelli describes a project planning meeting for agreeing on schedules and allocating future work. Work schedules and milestone charts, though, are only "fixed" ephemerally at each occasion of negotiation; and they will be revised at each subsequent meeting. In my own research, images that biologists and managers draw of themselves and each other—images often given to caricature in their self-agrandizement and denigration of the other—are images created in some specific context of argument and conflict. There is some point to the use of a particular symbolic representation at the particular time that someone chooses to use it. This contextualized meaning-in-process is like the concept of "point" discussed in the literary theory of narratology. Simply put, it says that stories have meaning, or "a point," because of the implied situation or larger context that the writer asks readers (or listeners) to imagine.[18] These situated meanings, one at a time, do not "add up to" some summary, general meaning or "account for" it (in ethnomethodological jargon). Rather, in our studies, they "point to" or, better, evoke some larger context in which they are situated and where they become meaningful. The point of argument between managers and executives (Chapter 5), or the point of negotiation between two engineering groups (Chapter 3), may well be the temporary fixing of the order of time: the schedule, the plan, the next thing to do. These "next things," these constructions, become meaningful in widening contexts—the contexts of what was done beforehand, what else is going on, and what will be done later.

THE MEDIA OF TEMPORAL ORDER

Times are key symbols around which multiple meanings coalesce, and they are embodied in different media like speech and writing, narrative accounts, and argumentative interchanges. There are a plethora of visual images—charts, schedules, plans, photographs, and video images—and time is also embodied in a variety of solid objects and machines, like particle detectors, solar panels, imaging equipment, biological substances, and laboratory apparatus. These condensed symbols of time provide a kind of "handle" with which its meaning can be grasped,[19] modified, and exchanged with others.

Larry Bucciarelli discusses the use of visual representations of time in the design process. As he points out, the imagery of "continuity" and the "flow of work" are central to a designer's perception of time. These include the "block diagrams" from standard textbooks (*vide* Figures 3-1 and 3-2) that represent the stages of a design process by geometrical shapes, and then link the shapes (stages) by arrows indicating the flow of action. This flow can go "forwards" or "backwards" (in loops), according to the logic of the process.

These images from the design theoreticians, however, are supplanted in practice by another set of visual tools propagated by the working managers of a design group or project as a way of exercising control over the process. These visual tools provide a square grid representing "tasks" versus "time" along two perpendicular axes. They list the division of labor (tasks) on a vertical axis, and juxtapose them with a linear, progressive time scale along a horizontal axis. The chart usually displays "time" from left to right, and has actual calendar dates as "milestones" or markers on the path to project completion. As images of future work and promises, these charts inject an element of fantasy in their overprecision and create a focus for negotiation between promises and accomplishments. Another charting device of managers, which also may or may not have a calibrated time axis, is the "Critical Path Method" (CPM) chart. This is a system for visualizing the relative importance of tasks in time. Stages of a process are represented like the geometric shapes of the logical flow diagrams and connected by arrows of time sequence. All arrows, in this case, point in the same (temporal) direction; and tasks are weighted by their ability to retard other concurrent or consequent tasks. The effect is an immensely complex and detailed image of the project as logical and fixed. Bucciarelli argues that all these models miss the point that actual design engineering is a flexible, negotiable process.

Any given timing chart, after any given session where it is drawn or revised, appears to represent the actual state of affairs—and plans for the future—of the group. It is a "captured" image of work flow, a symbol that conjures up an orderly past and—what's more—an orderly future.[20] The work of negotiation and interpretation disappears into the apparent fixity of the visual image. As this visualized future becomes past, however, the malleability and essential interpretability of time become evident in the continually repeated process of revising the chart. There is always a "new" present and a reworked future to be negotiated and redrawn into a new schedule. Similiarly, in biotechnology start-ups (Chapter 5), analogous representations of pasts and futures are the topic of ever-changing company plans, quarterly and annual reports, stock prospectuses, and investor presentations.

In commercial biotechnology development two prestigious and powerful groups of professionals come together to exploit contemporary advances in biology. Though they collaborate in company planning as founders, and then as business executives and senior scientists, the histories of their relationships are littered with tales of antagonism, conflict, and misunderstanding. Managers appear as short-term and biologists as long-term planners. Managers plot their research schedules as milestone or CPM charts, with clear beginnings, tasks, goals, and progress markers along the way. The biologists, by contrast, envision an infinitely branching tree of possibilities. They can imagine a distant goal, but they can't plot out a simple path through the vagaries of nature to reach it. At each step of the way, their craftwork experience suggests the next best choice from among several; but to chart a way clear through all contingencies is impossible. Each group, furthermore, connects their own style of planning to the "nature" of their professional work, be that the cultural construction of a scientific or an economic "reality." Each group perceives its stance as "realism," a proper orientation to the way things are. As in Barley's discussion, their very styles and patterns of approaching evidence and argument differ enough sometimes to impel them into anger and frustration. Unlike the circumstances of the doctors and technologists, though, there is no mechanical referee. Just as with Bucciarelli's engineers, there are only the documents, plans, and promises they produce together.[21]

Their native formulations of differences characterize conscious separation of scientists and executives and the maintenance of special group identities. Their self-images and caricatures of the other group are also ammunition in an ongoing struggle over who should guide or control company direction and research and development (R & D) policy. The spatial separation of the two groups in the firm and their frequent intra-group communication, as opposed to cross-group contacts, provides an environment where their images of each other can be honed to a cutting edge. These are often images of time, including caricatures of the developmental processes of lives and careers, on the one hand, and of the temporal patterns of work, on the other. Managers castigate their scientist colleagues for being "childlike" in their inability to define closure in research plans. What managers see as "immaturity"—not coming to completion—the scientists see as necessary, ongoing development.

Timing charts and schedules are not the only captured visual representations of time. Barley describes how x-ray images function as the primary articulation points between the highly discordant or "asynchronous" temporal orders of technicians and doctors. By contrast, new imaging technologies—particularly CT (computerized tomography) scanners—provide more closely integrated schedules of activity for the two groups. There, the examination and interpretation of the images themselves has become a joint activity. Radiologists and technicians tend to work in the same room, examining the CT images together, as the images are being produced.

For particle physicists (Chapter 2), the visual images

that are produced as photographic impressions of the collisons of elementary particles are crucial cultural artifacts. They are the singular locus where events in the non-relativistic time of experimental practice are captured and transformed into the "time-less" data with which physicists characterize the universe. Events in ephemeral time are stopped on film. Traces of contrast on a film are the concretized evidence of temporal motion, which is then reinterpreted as the atemporal, recurrent, replicable reality of elementary particles. These images, as the "produced" results of work, also validate physicists themselves and form the basis for their claims to honor, prestige, and financial resources.

Functioning in a more complex way than the "mechanical referee" for radiologists and doctors, the physicists' detector serves as a key symbol for the community of experimental particle physicists. Each detector is a condensed history of the research group's life together, and a mnemonic device through which the group can recount its history to itself. Traweek's analysis of the ways the detectors are both shaped by and, in turn, shape the lives and work of physicists makes clear that these machines are symbols in the sense raised in the first section of this chapter: they are embodiments of times and central social ordering devices for the physicists who build and use them.

Machines provide a medium for linking different temporal patterns of work and their consequences with the specific technical objects produced and used by the professional communities we examine. In Chapter 3, the solar energy panel is an emerging object in which temporal order is condensed, and then continuously restructured. In Chapter 4, the imaging technologies are already-given objects that articulate the structure of work processes and social relations. By comparing these two kinds of objects—one under construction and one in use—we can highlight the interconnected relationships among order, power, and knowledge in the social construction of time. The symbolization or articulation of temporal differences in machines provides a context for drawing together many of the issues raised earlier.

In the solar panel engineering design group,[22] there is a constant (if apparently gentle) tug-of-war that characterizes the relations among the design task groups. Their "object of attention" is gradually shaped out of a series of formal and informal interactions, meetings, and periods of individual work, through a kind of negotiation within the cultural bounds of engineering "common sense." The solar panel emerges from the interactive consultations and the flow and timing of work in the engineering design subgroups. Retrospectively, however, like emergent time, the panel is "simply there"; and the social process of construction recedes into invisibility.

The contrast with Barley's hospital professionals is striking. There, temporal uncertainties and asymmetries create conflict (as they also do in genetic engineering firms), while, among the design engineers, asymmetry and uncertainty are a positive, motive force. These groups' differing relationships to the technologies and artifacts that are central to their work provide a key to the difference. To hospital staff, the imaging machine (x-ray, ultrasound, or CT scanner)—the artifact—is largely a *fait accompli.* The requirements for running it effectively have been "standardized" by manufacturers, and then taught to the new users as a routine. They apprehend it as an externally created, relatively immutable presence around which work must be organized. The artifact appears to impose temporal order on the users. Technologies in use appear to impose an external temporal order; they structure time. Technologies under construction appear to engage all the task groups in an interactive, continuous process of negotiating their relationships, the flow of time in work, and the technical character of the object.

Hospital Radiology	*Design Engineering*
• Machines as given	• Object under construction
• Machine structures temporal order	• Machine structures social space
• Social space turns around the interpretation of products of the machine	• Temporal continuity turns around ongoing interpretations of and construction of the object
• Use of the new machines transforms an existing social order	• Constructing the object builds temporary social networks, reaffirms existing order

Let us extend this kind of comparison to another pair of examples from biology and particle physics. The machines and experimental apparatus in a typical wet biology lab, such as those of molecular biologists both in biotechnology firms and in academia,[23] are basically "given," in the same sense as imaging technologies in hospital labs. The biologists do not usually build their own apparatus; they tend to buy it "off the shelf," instead. Comparing these tools with the hand-crafted detectors of the experimental particle physicists suggests the following differences:

Laboratory Biology	*Particle Physics*
• Machines are bought off the shelf, "given"	• Detector is constructed, used, modified
• Machine output structures narrative	• Machine structures social life, careers, work, physicists' personae
• Machine output is "timeless" image of data: chart recorder traces, gels, numbers	• Machine is tool for constructing recurring time; its use structures calendrical (ephemeral) time
• Machines produce science, (and sometimes) biological substances	• Machine produces science, scientists, and nature
• Machine as transparent (disappearing) mediator between nature and evidence	• Machine resolves (and conceals) contradictory images of time

This final comparison highlights an important contrast. Biologists' machines disappear from their arguments, or are noted only peripherally. For biologists, the real focus of discovery is the biological substance, like TRF in Latour and Woolgar's ethnography. This substance, like a potential biological product for a genetic engineering company, becomes the object under construction, like the solar panel of Bucciarelli's engineers. The biologists' machines are mostly articulation points, like Barley's x-ray and ultrasound imaging devices, whose purpose is to produce an interpretable image. The biologists' craftwork practices, on the other hand, are more like the work of the solar panel design engineers. The biologists also invent, combine, create, and refine some new substance whose character is indeterminate until it is socially declared as "fixed" by some definitive scientific demonstration enshrined in a publication or patent.

As the particle physicists like to say of themselves, they have "more in common with each other as physicists, regardless of nationality or ethnicity, than they [do] with their next door neighbors, and, perhaps, even their families." This uniqueness challenges the unspoken assumptions that color traditional Western analyses of time. In a sense, we are trying to make our cases about making times by picking the hardest cases to argue—times embedded in the authoritative cultures and professional practices of science and technology, cultures whose mythological and monolithic images our studies help erode. If we have shown that the culture and praxis of advanced sciences and technologies do not reduce times to "time's arrow," then we have succeeded in calling the common wisdom into question.

NOTES

Acknowledgments: I thank the contributors to this collection for creating the occasion to write this essay; and I thank

Michael Ames (Temple University Press), Anna Hargreaves (Lotus Development Corp.), Ruth Linden (Brandeis), Renato Rosaldo (Stanford), and several anonymous reviewers for their careful critical readings of this introduction in its various incarnations.

1. All our case studies were conducted as ethnographic research, including extensive participant-observation, on-site visits, and interviews. Traweek and Dubinskas have doctorates in anthropology, Barley's doctorate is in organization studies with a sociological fieldwork orientation, and Bucciarelli's field research was designed on an anthropological model.

2. Our essays contrast with much of the work on corporate and organizational cultures by focusing on professional communities, who share more patterns of culture within their own groups than they do across their organizations as a whole. This is not to say that organizations like the biotechnology firms, hospitals, laboratories, or energy companies never engender their own common cultural variants that are shared by most members; but this should be an issue for investigation, not presumption.

3. I describe a similar process in the historical transformation of dominant temporal models in conjunction with the spread of mechanization of agriculture in rural Yugoslavia in the 1950's and 1960's in my essay "Time and Style" (Dubinskas, 1980). I examine radical changes in musical stylistics in village singing, a form of expressive culture where temporal patterning is an important indicator of generational differences among singers. These musical changes are congruent with other transformations in village farmers' lives, especially around formal schooling and agricultural work.

4. This is a paraphrase of "time is only a shadow cast by motion," quoted by Shigeo Shingo in a 1986 seminar on manufacturing productivity. The original source is cited as F. Gilbreth's classic "time and motion" studies (Gilbreth, 1914).

5. See Rose (1975) for a discussion of the history of Taylor's influence and ideas. For the original works, see Taylor (1911) and Gilbreth (1914).

6. I use the plural "biotechnologies" because the fundamental technical revolutions upon which these small companies' work is based are relatively independent. Genetic engineering or "gene splicing" deals with the rearrangement of DNA. Monoclonal

antibody technologies deal with the isolation and production of identical immunological molecules through a different process. These two technologies helped spawn most of the new "biotech" firms of the early 1980's.

7. For a critique of this implicit rhetoric of ethnography, see the various works in Clifford and Marcus' collection *Writing Culture* (1986) and Marcus and Fischer's *Anthropology as Cultural Critique* (1986).

8. Latour and Woolgar's field study (1979) at the Salk Institute in La Jolla, California, is on work in the laboratory of Prof. Roger Guillemin on establishing the existence of "tryptophan releasing factor/hormone" (TRF).

9. See also Woolgar (n.d.) on the interpretation of the trace of a recording pen on chart paper as an example of the practical construction of a scientific fact.

10. The paper by Harold Garfinkel, Michael Lynch, and Eric Livingston (1981) is an ethnomethodological analysis of audio tapes that were accidently and felicitously made during the astronomical discovery of the optical pulsar.

11. Bruno Latour and Steve Woolgar's classic, *Laboratory Life: The Social Construction of Scientific Facts* (1979), is the first modern ethnographic study of advanced scientific laboratory practice; and it was conducted by Latour at the Salk Institute in Guillemin's neuroendocrinology laboratory. Their discussion of "circumstances" (or context) is somewhat broader than other micro-sociological work from a more strictly ethnomethodological approach. They treat the economics of reputation, built through professional networks, as crucial to the abilities of scientists to establish facts. This work was followed in publication by Karin Knorr-Cetina's *The Manufacture of Knowledge: An Essay on the Constructivist and Contextual Nature of Science* (1981), based on a study of work in a biology laboratory at the University of California–Berkeley. An introduction to further literature in the constructivist micro-sociological genre may be found in the review article by Woolgar (1982) and in the collection by Knorr-Cetina and Mulkay, eds. (1983). Pinch and Bijker (1984) extend the constructivist arguments to artifacts and technologies in their review and programmatic essay, "The Social Construction of Facts and Artifacts: Or How the Sociology of Science and the Sociology of Technology Might Benefit Each

Other." The journal *Social Studies of Science* (London: Sage Press) regularly presents current work from this school.

12. In fact, we do not treat in detail the widest possible systems that provide relevant contexts for understanding our subjects. For instance, the realms of national and international politics and economics, which certainly help shape the social environments and cultures of our subjects, are addressed only *passim.* Our foci remain largely internal to organizations and to professional communities; but, in each case, their "local" politics and economies are still central to our arguments.

13. See Benedict (1934) for the source discussion in cultural anthropology of "pattern" and "configuration" as analytical categories. Benedict uses "configuration" to describe the interrelations among more localized cultural patterns. Configuration is thus a kind of meta-pattern, but one that does not imply strictly formal, generative, or rule-bound connectivity.

14. See especially chap. 2, "Oecology," and chap. 3, "Time and Space" (Evans-Pritchard, 1940). Traweek notes several other important ethnographic studies outside of Western cultures in note 2 of Chapter 2, below.

15. I have participated as a staff researcher in a project on new car development in the worldwide automobile industry, sponsored by the Harvard Business School and directed by Prof. Kim B. Clark of the Production and Operations Management Group. This project, ongoing since late 1984, includes field research, with company collaboration, in eight Japanese, three U.S., and nine European auto companies on the management and performance of new car projects. My own work includes managing the European research and conducting some of the U.S. research, as well as participating in the analysis and interpretation of data from all sites.

16. See "Notes on the Balinese Cockfight" (Geertz, 1973b).

17. We employ ethnographic research approaches of observation, interviews, and engagement, as well as other traditional techniques of data collection. These other sources include a wide variety of written and/or published documents by our subjects as well as other social scientific and historical accounts. Many of the "objects" at issue in our studies, like milestone charts, business plans, photo-images, and research protocols, are also readily accessible to us for examination.

18. This is a formulation similar to the stance that Umberto Eco takes in *The Role of the Reader* (1979). In Eco, the writer is the creator or implied tutor of readers, and hence the more powerful actor in the narrative process. The writer establishes the context. A more interactive model of fictional narrative is proposed by Ross Chambers. In *Story and Situation* (1984), he describes a situated meaning as "the perception of a relationship between discourse and its context" (p. 3). He further describes narrative "as a transactional phenomenon. Transactional in that it mediates exchanges that produce historical change, it is transactional, too, in that this functioning is itself dependent on an initial contract, an understanding between the participants in the exchange as to the purposes served by the narrative function, its 'point.'. . . It is this contractual agreement as to point that assigns meaningfulness to the discourse" (p. 8). Point is thus a shared context of meaning, rather than an isolated instance or singularly generated one. The role of narrative or, more generally, any symbolic action as a mediating function in social transactions is addressed in more detail in the discussions on social order and conflict and on the symbolization and embodiment of times.

19. It is interesting to note that "grasping" an image—understanding or "getting" it—has a paired expression that places the force of action within the symbol, not just the interpreter. This is the sense of a "gripping" image. This reversibility is implied in my descriptions of the self-other characterizations of biologists and managers, and it is explicit in Traweek's treatment of physicists' anxieties.

20. Bourdieu comments extensively on this "capturing" of temporal structure (1977, pp. 6–9).

21. The biological substances and organisms with which they work, however, may play this role. Bruno Latour argues like this for the role of the microbe in Pasteur's career (Latour, 1984).

22. Larry Bucciarelli has conducted his research both as an ethnographic observer and also as a consulting engineer. He was engaged to participate in the design group as an expert contributor on both the electrical network and structural design of the solar panel.

23. I draw here on my own field research and my five years' experience as a laboratory research assistant in biochemistry and

biophysics. The details of laboratory practice are also represented in Latour and Woolgar's *Laboratory Life: The Social Construction of Scientific Facts* (1979). For other relevant sources, see note 11, above.

REFERENCES

Benedict, Ruth. 1934. *Patterns of Culture.* Boston: Houghton Mifflin.

Bohannon, Paul. 1953. "Concepts of Time Among the Tiv of Nigeria." *Southwest Journal of Anthropology* 9, no. 3 (autumn).

Bourdieu, Pierre. 1977. *Outline of a Theory of Practice.* Cambridge, U.K.: Cambridge University Press.

Chambers, Ross. 1984. *Story and Situation: Narrative Seduction and the Power of Fiction. Theory and History of Literature,* ed. Wlad Godzich and Jochen Schulte-Sasse, vol. 12. Minneapolis: University of Minnesota Press.

Clifford, James, and George E. Marcus. 1986. *Writing Culture: The Poetics and Politics of Ethnography.* Berkeley: University of California Press.

Dubinskas, Frank A. 1980. "Time and Style in Slavonian Folk Music: Elements in a Culture History of Temporal Consciousness." Paper presented at American Society for Ethnohistory annual meeting, Oct. 25, San Francisco.

Eco, Umberto. 1979. *The Role of the Reader: Explorations in the Semiotics of Texts.* Bloomington: Indiana University Press.

Evans-Pritchard, E. E. 1940. *The Nuer: A Description of the Modes of Livelihood and Political Institutions of a Nilotic People.* London: Oxford University Press.

Garfinkel, Harold, Michael Lynch, and Eric Livingston. 1981. "The Work of a Discovering Science Construed with Materials from the Optically Discovered Pulsar." *Philosophy of the Social Sciences* 11 (June).

Geertz, Clifford. 1973a. *The Interpretation of Cultures.* New York: Basic Books.

———. 1973b. "Notes on the Balinese Cockfight." In *The Interpretation of Cultures.* New York: Basic Books.

———. 1973c. "Person, Time, and Conduct in Bali." In *The Interpretation of Cultures.* New York: Basic Books.

———. 1973d. "Religion as a Cultural System." In Geertz, *The Interpretation of Cultures.* New York: Basic Books.

———. 1973e. "Thick Description: Toward an Interpretive Theory of Culture." In Geertz, *The Interpretation of Cultures.* New York: Basic Books.

Giddens, Anthony. 1984. *The Constitution of Society: Outline of the Theory of Structuration.* Cambridge, U.K.: Polity Press.

Gilbreth, F. 1914. *Primer of Scientific Management.* Easton, Md.: Hive.

Gould, Stephen Jay. 1987. *Time's Arrow, Time's Cycle: Myth and Metaphor in the Discovery of Geological Time.* Cambridge, Mass.: Harvard University Press.

Hall, Edward T. 1984. *The Dance of Life: The Other Dimension of Time.* Garden City, N.Y.: Anchor Press/Doubleday.

Knorr-Cetina, Karin. 1981. *The Manufacture of Knowledge: An Essay on the Constructivist and Contextual Nature of Science.* Oxford, U.K.: Pergamon Press.

Knorr-Cetina, Karin, and Michael Mulkay, eds. 1983. *Science Observed: Perspectives on the Society Study of Science.* London: Sage Press.

Kuhn, Thomas S. 1962. *The Structure of Scientific Revolutions.* Chicago: University of Chicago Press.

Landes, David S. 1983. *Revolution in Time: Clocks and the Making of the Modern World.* Cambridge, Mass.: Harvard University Press.

Latour, Bruno. 1984. *Les Microbes: guerre et paix suivi de irréductions.* Paris: A.-M. Métailié.

Latour, Bruno, and Steve Woolgar. 1979. *Laboratory Life: The Social Construction of Scientific Facts.* Beverly Hills, Calif.: Publications.

Marcus, George E., and Michael Fischer. 1986. *Anthropology as Cultural Critique.* Chicago: University Chicago Press.

Pinch, Trevor J., and Weibe E. Bijker. 1984. "The Social Construction of Facts and Artefacts: Or How the Sociology of Science and the Sociology of Technology Might Benefit Each Other." *Social Studies of Science* 14, no. 3.

Rose, Michael. 1975. *Industrial Behavior: Theoretical Development Since Taylor.* Harmondsworth, U.K.: Penguin Books.

Taylor, Frederick W. 1911. *The Principles of Scientific Management.* 1967 ed. New York: W. W. Norton.

Tedlock, Barbara. 1981. *Time and the Highland Maya.* Albuquerque: University of New Mexico Press.

Thompson, E. P. 1967. "Time, Work-Discipline, and Industrial Capitalism." *Past and Present* 38.

Woolgar, Steve. 1982. "Laboratory Studies: A Comment on the State of the Art." *Social Studies of Science* 12, no. 4.

———. n.d. "Time and Records in Researcher Interaction: Some Ways of Making Out What Is Happening in Experimental Science." Unpublished MS.

Zarubavel, Eviator. 1979. *Patterns of Time in Hospital Life.* Chicago: University of Chicago Press.

———. 1981. *Hidden Rhythms: Schedules and Calendars in Social Life.* Chicago: University of Chicago Press.

CHAPTER 2

Discovering Machines: Nature in the Age of Its Mechanical Reproduction

Sharon Traweek

THIS CHAPTER[1] addresses the relationship in scientific research among the production of a research community, the production of scientists, the production of knowledge, and the production of research equipment. Anthropologically speaking, I am examining the relationship in the high energy physics community among their social organization, their developmental cycle, their cosmology, and their material culture. It is time that is at the nexus of this relationship: structuring time organizes the high energy physics community. Knowing those time constraints and living within them structures the physicists themselves.

In my study I have adapted an hypothesis that has been developed in many ethnographies over the last several decades, which is that a culture's ideas about time and space also would be manifest in the domain of action. That is, their ideas about time and space would structure social interaction and, on the other hand, the spatial and temporal organization of human activity would be restated in their conceptual formulations about time and space.[2] I have argued that this is true for the physicists. That is to say, that their physical theories about time and space construct and organize their social reality and their social reality is restated in the domain of these physical theories.

I am discussing these physicists' production of discovery, detectors, laboratories, and novices as the production of representations of science and as the production of the means of representation of nature.[3] In my emphasis on detectors as a crucial exemplar of that practice I am arguing that scientific instrumentation plays a critical role in the production of scientific knowledge. Furthermore I am asserting that the usual distinctions (and presumed relations) made between science and engineering, between theory and technology, between knowledge and skill, must be reconsidered. The detectors in high energy physics are produced according to evolving, community-determined criteria, and they each also display a distinctive style in the execution of those criteria. The study of the production of artifacts as cultural performance, as material culture, has long been a practice of anthropology. That practice of producing representations is the collective activity of a community, its culture, and its cultural practice.[4]

I take a society's culture to be their representation of their world to themselves, and I study the process by which this community generates, maintains, and modifies that representation: their construction of common sense, if you will. The representation that I call culture is, of course, composed by many, specific diurnal practices that might be called micro-representations. It is from observation of and participation in these practices that I construct an interpretation of the recursions and correspondences among this society's diverse representational practices, and then identify the characteristic type of correspondence that is distinctive to that group, calling it culture. It is necessary to remember that the representational practice of generating descriptions of peoples' culture is itself characteristic of the society called anthropology, operating according to our own canons of narrativity and allusive reference. But that is another tale. What follows is one of my constructions of how the high energy physicists construct their world and represent it to themselves as free of their own agency: a thick description of the culture of objectivity, the culture of no culture, a culture that knows it takes the world neat.[5]

For the past ten years I have been studying high energy physicists who, in turn, study something called particle physics or quantum electrodynamics (QED), which is the analysis of the basic constituents of matter and their relations. Their research is conducted with a very few, very expensive, and very large machines called accelerators in which certain particles are accelerated to very high energies and then caused to collide with other particles. Research devices called detectors are used to examine the debris from these collisions. Detectors produce data about collisions and that data is used to affirm or reject or even confound existing theories.

There are about 800 to 1,000 active high energy physicists in the world, concentrated in the northern hemisphere. About 3 percent of them are women. About 300 to 400 of these people know each other well and the others want to. About half are called theorists; they work alone or in small, short-lived groups of two or three people at blackboards. The other half are called experimentalists, and they work with the machines in long-lived groups of about twenty people. The experimentalists and theorists both work together, but they keep a stylized distance from each other.

There are about ten important accelerators, located in the United States, Western Europe, the U.S.S.R., and Japan. My primary fieldwork has been conducted at three national laboratories over a period of five years: Stanford Linear Accelerator (SLAC) near San Francisco, Tsukuba Laboratory for High Energy Physics (Ko-Enerugie butsurigaku Kenkyusho, or KEK) near Tokyo, and Fermi National Accelerator Laboratory (Fermilab) near Chicago. I also have visited other laboratories, such as CERN (European Laboratory for Particle Physics) in Geneva and DESY

(Deutsches Elektronen-Synchrotron) near Hamburg, as well as many university physics departments.

I began by explaining to the high energy physicists that I wanted to study the culture of particle physics and I also wanted to show how the cultural environment that surrounds the laboratory, whether American or Japanese or whatever, might influence that local culture of particle physics. Many of them agreed that there was a culture of particle physics. They felt that they had more in common with each other as physicists, regardless of nationality or ethnicity, than they did with their next door neighbors, and, perhaps, even their families. However, the idea of there being a culture from the outside impinging on them was very annoying to them. One physicist in particular said, "Look, Sharon. Culture is like a Poisson distribution. You have to understand that scientists are drawn from out here in the tail of the distribution where cultures have very little impact." In other words, he saw culture and reason as standing in inverse relation to each other. The physicists often use scientific analogies for talking about human behavior. They use quantifiable terms to discuss activities that cannot be quantified, or are very difficult to quantify. They would prefer that matters social be quantifiable. They are also trying to transmit an essential point about human relations. In a Poisson distribution one presupposes discrete, unrelated entities and that is exactly the way that they see the human population: each person is unique; each has a certain, probably genetically determined, capacity for rational thought, and as a consequence each is more or less receptive to the cultural biases that might be imprinted on that person. They see scientific education as a matter of stripping away these biases, much more than trying to embed reason in anybody, because they believe that cannot be done.

How should we look at scientific machines? Most analysts of science and most scientists claim that it is the size and complexity of the big machines that determine the organization of scientific work, perhaps even determining the research itself. I am suggesting a careful interpretation of how scientific machines shape scientific practice and how scientists regard the machines that they are building—and rebuilding—in perpetuity. I am asking what these machines are and proposing these answers: they are machines for the production of nature, the production of discoveries, and the production of scientists; they are also machines that display nature, discovery, and scientific genius.

DETECTORS

In order to understand the place of machines in high energy physics, it is necessary to recapitulate the rudiments of the research process in high energy physics. The accelerator is a device used for impelling particles to very high energies, which then are directed at targets. The collision of the target particles with the accelerated particles generates energy; this energy mostly takes the form of new particles. (There is a residue of energy in the form of radiation. Concrete blocks are used to shield people from that radiation in areas where it is particularly strong.) Targets (collison areas) are surrounded by diverse devices designed to record traces of the new particles, so that they can be identified. The target plus the recording device and computing system is called a detector.

At any laboratory there are many detectors near the accelerator. Since the accelerated particles are clustered in discrete bunches, those bunches can be delivered (by means of the magnets in the "switchyard" bending their path) to any of the various detectors. One detector may be able to utilize 60 pulses per second; another, 10; yet another, 30. The full load of the accelerator beam (120, 180, or 360 pulses per second, for example) is divided in this way to the several experiments that are being conducted "simultaneously."

Each group of experimentalists conceives, constructs, maintains and develops its own detector, each of which is distinctive and serves as the signature of the group. These machines are at the heart of the research activity of particle physicists. Discovering a new way to detect the presence of elementary particles by recording their traces can bring great honor and influence. Especially prized is a machine that combines sophistication and subtlety in its collection and resolution of data with elegance of construction. If it were to run perfectly at all times, however, a detector probably would be considered either obsolete or not daring enough in conception. The place of detectors in high energy physics contrasts sharply with Bruno Latour and Steve Woolgar's study of an endocrinology lab (1979), in which the authors characterize the role of detector as "black boxes" or "reified theory." That kind of detector is not produced by research scientists but mass produced by manufacturing firms. The design of those detectors incorporates theories and laboratory practices so widely accepted that their validity has not been questioned for many years and perhaps many decades.

Most high energy physics detectors share certain characteristics. They establish a highly sensitive medium through which the particles generated by the interaction of the accelerated beam and the target materials are directed. That medium is disturbed by the activity of the particles; the disturbance of the medium is then meticulously measured. The record of the pattern of disturbances is then analyzed in order to deduce the presumed properties of the particles that caused the disturbance. Subtle but highly significant differences between detectors of the same type are based upon how sensitive the initial environment is, how effectively it can be controlled, how differentiated the disturbance of the medium can become, and how carefully the disturbance can be calibrated. Physicists also want to know how frequently this process can be repeated so as to accumulate a very large number of events during each experiment. This will enhance the validity of any conclusions they may reach and increase the probability that a rare event will be "seen." Furthermore, other aspects of the ambient environment besides the particles generated by the controlled collision do disturb the medium and trigger the measuring and recording equipment. These aspects are known as "noise." A fine detector should enable the physicists to identify and measure within an acceptable range the amount of noise generated in any specific experiment.

A highly sensitive medium that is capable of finely differentiated disturbance associated with a system that can measure those minute fluctuations is a delicate piece of machinery, quite vulnerable to high levels of noise production, some of which will be unpredictable. Experimentalists strive to maximize sensitivity and speed, while minimizing noise, especially unpredictable noise. In the view of high energy physicists, a mass-produced detector has an unacceptably large margin of conservatism built into the machine. Once one of their own detectors begins to work with great regularity and predictability, it becomes a candidate for this mass production and potential distribution to scientists in other fields. Such a detector is considered obsolete by high energy physicists. A detector can be discarded (or never even built) for another reason: cost. The price of achieving the sensitivity, calibrating the noise, or "reading" the data can become prohibitive.

Given these common features and problems with detectors in high energy physics, differences among types of detectors can be seen as maximizing one of the component variables (sensitivity in identifying the presence of particles, speed of data collection, capacity of distinguishing noise, and mode of data analysis). Bubble chambers probably provide the most elaborate data on particle behavior;

however, they collect data much more slowly than other detectors, and since the data are recorded on film rather than in computers, analysis is lengthy and costly. At the other extreme are "counters" that merely signal that some particle has passed through a sensitive grid of wires; this information can be recorded immediately in computers. Analysis of these data can begin while the experiment is still running ("on-line data analysis," in "real time"), enabling the experimentalists to know the quality of data they are gathering and altering the experiment to accommodate any problems. Detector development is devoted to finding new ways to collect quickly complex data that can be recorded directly into computers for on-line data analysis.

There have been nine detectors in the SLAC research yard: three bubble chambers, a spark chamber, a streamer chamber, a large aperture solenoid spectrometer, a group of three spectrometers, and two detectors associated with a colliding beam facility. The bubble, spark, and streamer chambers represent refinements of decades-old innovations. The others represent newer developments in the conception of machines designed to detect traces of nature.

REPRODUCING NATURE: THREE DETECTORS IN THE RESEARCH YARD

Over the past ten years I have discovered that most non-scientists believe that laboratories are extremely clean, meticulously tidy places where people in immaculate white coats engage in activities that require minute, precise movements, and that scientists work alone in silence. High energy physics laboratories are not like that. The research yard at SLAC resembles nothing so much as a big, busy, messy manufacturing business. At one end is a monolithic concrete structure, seven stories high. A few corrugated metal two-story buildings and several sheds stand in haphazard relation to each other. Scattered among them are many concrete blocks (one by two by three meters), large cable spools (one and two meters in diameter), stacked lumber, huge cranes, some toilet sheds, and clusters of automobiles. The entire area is paved. The people wear work shoes or "running" shoes, jeans or khaki pants, and T-shirts or work shirts. All look a little dirty. A few wear safety helmets. Getting out of the car and walking around, one is confronted by obstacles in one's path in almost every direction. These raised burrows turn out to be makeshift housing for the electrical cables, thick as a fist, which connect the detectors to power sources and to computers. One crosses these by means of little ladders, much like the stiles used for crossing fences in rural areas.

Walking toward the massive concrete structure, one notices that along one side its huge walls move on rails like barn doors. The thick concrete provides protection against the radiation produced in the experiments conducted inside. Like the accelerator and beam switchyard no one is allowed in the building during an experiment: the research devices are controlled remotely during a typical "run" of several weeks.

Experiments here can make use of the maximum energy of the beam and up to 320 of the beam's 360 pulses per second. When the incoming accelerated electrons collide with target protons the electron is said to "scatter" and the proton to "recoil" ("elastic electron scattering"); in some interactions new particles are also generated ("inelastic electron scattering"). In some experiments the electron beam is directed at a preliminary target, the products of which (photons) are used as a "secondary" beam to collide with the usual target photons. In this case (photon-proton interactions) the proton "recoils" and new particles are also generated ("photoproduction"; i.e., production of new particles by means of a photon beam). The detectors in End Station A (three spectrometers) are designed to measure the amount of electrons scattered, new particles produced, and the recoil of the proton.

The usual access to End Station A (when the walls are not "open") is through a long corridor that opens onto the vast 2,500-square-meter interior space of the building. Walking into the room, one notices three very large concrete boxes, each with a thick protruding limb. These limbs all join in a single point. At the central point is the target, a drum-sized metal container. Access to the target is by a catwalk over the large open pit situated underneath the target. In that pit is a large pivot made from a 16-inch World War II Navy surplus battleship gun. Each of the three big concrete boxes is attached to that pivot by different-looking limbs. Embedded in those protuberances are paths for scattered electrons and new particles: those are surrounded by a series of magnets that bend and focus the particles—procedures that will enable the detectors in the large concrete boxes to begin reconstructing the momentum and position of the particles at the moment of collision. The two longer limbs are about 2 meters wide and high; one is about 50 meters long, the other about 25 meters. The concrete boxes appended to those limbs open to reveal shifting configurations of detectors. Originally simple counters were arrayed in the boxes. By 1971, these were supplemented by multiwire proportional chambers, so that the data about the scattered and newly produced particles could be further refined. This added information also makes possible better detection and rejection of data from unwanted sources. The longest spectrometer, called the 20 GeV, is 50 meters long and weighs 2,000 tons; the shorter 8 GeV is 25 meters long and weighs 1,000 tons. The third and shortest spectrometer, the 1.6 GeV, weighs 575 tons ("SLAC and MIT Collaboration Studies Proton Structure," 1970; Kociol, 1970; Oxley, 1971; Kirk, 1975b). The larger spectrometers analyze scattered and newly produced particles; the smallest spectrometer analyzes the recoiled protons.

Although these three spectrometers are very limited in

the particle characteristics they can identify, they have two important advantages: first, an extremely large number of events can be analyzed at 320 pulses per second; second, all three spectrometers can be moved. They are all anchored at that central pivot beneath the target so that they can, together, rotate through 165 degrees around the target. Rather than creating a three-dimensional stationary detector, this detector's designers conceived of a movable detection system. The spectrometers roll on concentric rails around the pivot. Large electrical cables spew out of the boxes' detectors, and are carried alongside the magnets, down the arms of the spectrometers, to the pivot in the pit. From there, they are guided along the floor and up the wall, where they enter the Counting House, End Station A's control room and computer facility.

The pervasive grey of the concrete at End Station A and the large, lumbering detector boxes with their supporting equipment reaching toward the pivot pit always look to me like great mechanical elephants with their trunks plumbing a watering hole. The room and the spectrometers appear massive, mechanical, and clumsy. It is difficult to remember that it is the fastest detector at SLAC and the site of more than one experiment of major theoretical consequence. The scattering experiments have served to arbitrate between theories on the internal structure of the proton. The design and integration of a polarized electron beam source (called "PEGGY") and a target with polarized protons at End Station A has enabled experimentalists to design a study that concluded that time reversal invariance is maintained, in fact, at the microscopic level.

There are two groups that oversee End Station A, one primarily devoted to research, the other committed to developing and maintaining the equipment. The research group leader is Canadian; he is a powerful figure in the high energy physics community, with strong views on the current state of the field and its future.

Nearby is another kind of detector, the Large Aperture Solenoid Spectrometer, known by its acronym, LASS. The name was bestowed by the Scottish leader of the group that has designed, constructed, operated, and researched with this device. Particle activity in the detectors called bubble chambers is detected three dimensionally because the medium in which the initial collision between accelerated particles and target particles occurs is also the medium of detection. The difficulty with bubble chambers, however, is their slow rate of data collection and the expense of analyzing data. The hybrid systems minimize these problems, but do not eliminate them. LASS represents another approach to these issues (Kociol, 1971; Oxley, 1972).

In an earlier manifestation of this detector, different kinds of counters and spark chambers (Cerenkov counters, scintillation counters, multiwire proportional chambers) were aligned behind a magnet. As with the bubble chambers, the purpose of the magnets is to alter the path of charged particles; since the strength of the magnetic field can be controlled, the particle path's angle of curvature and speed (as recorded by the various counters and chambers) is used as a measure of that particle's mass and lifetime. Several characteristics of those fast-moving particles that emerged in a "downstream" direction from the initial collision could be identified. This detector was called the neutral K meson (K^0) wire chamber facility ("the K-zero"), which referred to some of the machinery and the altered beam that it utilized. The group leader was very proud of the degree of resolution achieved in the multiwire proportional chambers, originally developed at CERN by Charpak. In the grid used to detect the presence of particles, wires were placed particularly close together, but they avoided having the whole system, in its sensitive state, trigger itself. Each wire "constricted" when "fired" by a passing particle; when that constriction passed (at a predictable rate) to the end of the wire at the edge of the metal frame, the computer was signaled. In this way, a great amount of very precise information was collected. He noted to me that the construction of these magnetorestrictive spark chambers and the Cerenkov counters (which identify particles by their characteristic light emissions) included techniques he had developed as a young physicist. He believed that his sense of good experimental work had been learned in his teachers' labs in Scotland; he was inspired too by the examples of even earlier physicists' experimental equipment that were displayed in the dining hall of his university laboratory. Scotland has an important history of empirical scientific research (Olson, 1975). This physicist knew that history through those devices he had observed in the dining hall and his teachers' stories, not through any study of written histories.

The K-zero was incorporated into LASS, much as the hydraulic drive system from an 82-inch bubble chamber was incorporated into later bubble chambers built by that group. The added features were designed to identify the slower-moving particles that had emerged from the initial collision at wider angles than the K-zero could detect. The new detector's target is surrounded by concentric sleeves of detecting devices. Immediately downstream are four superconducting magnets interspersed with more detecting equipment. There the fast forward-moving particles enter what is essentially the old detector: yet another magnetic field, followed by the two different kinds of counters and the spark chambers. This diverse array of magnets and detecting devices means that a great deal of information can be gathered about each particle path, even if it is not a three-dimensional detector. All the information is recorded by computers, and complex data analysis can be conducted during an experiment. Considerable computing power is required for operating the detector, data collection, and analysis, especially with the speed at which LASS is designed to operate: while the 82-inch chamber recorded 24.3

million events in five and a half years, LASS can record 100 million events in one year. These massive computing requirements eventually led to the acquisition of a major new computer for SLAC, the IBM 3081.

This machine has been designed to investigate interactions analogous to those explored at End Station A. In their scattering experiments, accelerated electrons collided with target protons generating new particles and causing the electron and proton to recoil. LASS would be able to study the same scattering process among different kinds of particles. In particular, LASS would analyze hadron-hadron interactions. Electrons have charge, which means that they belong to a class of particles interacting by means of electromagnetic force. Hadrons are particles that interact by means of the strong nuclear force.

LASS is about twenty meters long, three meters wide, and three to five meters high. Each of the components, aligned in series, has a distinctive configuration. The four separate three-meter-diameter coils of the superconducting magnet are smooth, thick, and bright. Sandwiched between these coils are the multiwire proportional chambers; almost all that is visible of them are their long electrical cables. This whole section rests on a system of rails and jacks so that it can be aligned and dismantled easily. All this is straddled by a long-legged rigging on which sits the refrigeration system for cooling the magnets. The rigging looks like a giant insect embracing a metal and plastic caterpillar.

Following this display is the conventional magnet, embedded in a one-meter-deep, three-by-four meter casing. Its coil is wrapped into a configuration that looks like giant lips. Behind this mouth are the 12 spark chambers, hung on their metal frames with masses of electrical wiring laced around and through the frames. In this context, the closely spaced chambers resemble a bellows-like alimentary canal. At the end of the entire tract is a dark, five-meters-deep, three-by-four-meter box containing the Cerenkov counters, which absorb the final data to be extracted from the event by analyzing the light emitted by the particles. Hovering over all this near the ceiling and moving on railings placed at the top of two opposing walls of the building is a 7.5-ton-capacity crane. The crane is used to dismantle and rearrange the components of the detector.

LASS took several years to design and construct; the final stages included many setbacks, especially in the magnets. During this time, doubts were raised at the lab and in the larger physics community as a whole about the viability of the detector, and whether the physics it was designed to do would prove worth the resources that had been allocated to it. Some physicists thought that the design was too precise and refined to be built; others thought the physics to be advanced but the detector mundane. These doubts, I had to realize, beset any project, especially once it is funded. Part of the responsibility of a group leader is to defend the group's project (and budget) against all efforts to diminish it. This group's leader is a very forceful member of the international high energy physics community.

Shortly after the LASS magnets began operating, a party was held in its honor. The research group played host in the building housing LASS to all who had contributed, directly or indirectly, to the existence of the detector. The party for LASS included huge kegs of beer; the group was known to gather often at a nearby tavern to discuss physics and their LASS. Holding a half-quart paper cup of beer in one hand, each group member could be seen pointing to parts of LASS and talking animatedly, or sitting in the control room bringing up interesting graphics on the CRTs about LASS's sensitive self-monitoring system.

North of LASS and beyond the bulwark of End Station A is a facility that almost failed to be built. It is called SPEAR, an acronym for Stanford Positron Electron Asymmetric Rings. The original design called for two pear-shaped rings; the final design has one symmetric ring, but the name remains. SPEAR is one of a growing number of "colliding beam" facilities at high energy physics laboratories. A colliding beam causes accelerated beams of particles to collide with each other, thereby doubling the center of mass energy available at the moment of collision. These collisions between an accelerated beam and a stationary target have a much reduced interaction energy, due to the energy taken up by the recoiling action of the original particles. For example, if each beam at SPEAR had an energy of 4.5 GeV (billion electron volts), the center of mass energy available at the moment of collision would be 9 GeV. It would take a conventional accelerator capable of accelerating a beam to 50 GeV to match this center of mass energy, and the numbers rise with the square root. The problem is especially serious in the case of electron accelerators such as SLAC, because of special characteristics of electrons at very high energies. A significant problem with colliding beams, however, is that the intersecting accelerated particles actually collide much less often than in the conventional configuration of beams colliding with stationary targets because the beam is less dense than a target (Kirk, 1974).

A special feature of colliding beams is that usually one beam is composed of the anti-particle of the other beam. For example, at SPEAR, one beam is composed of electrons from the linear accelerator. The other beam, also from the accelerator, is composed of "anti-electrons" known as positrons. An anti-particle has *most* characteristics in common with its particle except its electrical charge, which is reversed. Hence a positron is a positively charged electron. Positrons are generated in the accelerator when a partially accelerated beam of electrons strikes a tungsten or copper target inserted about one-third of the way down the two-mile accelerator. The resulting radiation (gamma rays), striking another target, generates electrons and positrons in pairs (pair production). The electrons are deflected by means of magnets, leaving a beam of positrons which then are accelerated the remaining length of the accelerator, (Kirk, 1975a). Colliding beam facilities use particle and anti-particle beams because of a special characteristic of particle and anti-particle interactions: both particles are "annihilated," with only energy remaining. This energy then reforms into new particles with no new or colliding particles to take energy as they recoil from their collision (as in the experiments at End Station A).

The first colliding beam facility was built at the Stanford University High Energy Physics Laboratory (HEPL) in the late 1950's and early 1960's. The leader of the group that designed, constructed, maintains, and does experiments with SPEAR was a member of the Stanford-Princeton team that built the HEPL ring. He submitted his first proposal for a high energy colliding beam facility at SLAC in 1964 (he had joined SLAC in 1963). That and subsequent proposals were rejected until 1970, when a much more modest one was accepted. Actually, SPEAR was not funded through the usual channels. The Atomic Energy Commission, the federal agency that then funded particle physics research in the United States, merely allowed SLAC to shift 5 million dollars from other projects in order to finance the construction of SPEAR. It was not clear at the time how successful colliding beam facilities were to become.

The man who led the SPEAR group had done important work in the 1950's on electron-positron interactions, establishing that the current analysis of the electromagnetic force was correct at extremely short distances (10^{-13} centimeters). He had decided he wanted to study particles that interacted by means of the strong nuclear force (hadrons). He has said that "it seemed to me that the electron-positron system, which allowed one to produce these particles in a particularly simple initial state, was the right way to do it. . . . That was the beginning of the long struggle to obtain funding for the device" (Richter, 1976).

In spite of the precariousness of funding worldwide for high energy physics since the early 1970's and the difficulty of funding SPEAR, a report circulated in 1972 at SLAC projected the following timetable: "In 1972—conceptual design and physics studies: 1973— begin detailed engineering studies and writing of proposals; 1974—submit official proposal to the AEC, asking authorization for construction to begin in FY 1976. With such a timetable, a PEP accelerator would be operating about 1980" ("LBL, SLAC Designing Colliding Beam Accelerator," 1972). With small perturbations, this timetable's predictions proved true. It is typical for new projects to enter the design stage immediately after other projects have been funded so that there is a continuous cycle of ideas seeking funding. Of course, many projects never make it beyond the design and proposal stage.

The main approach to SPEAR is by a narrow bridge over the ring, which is about 75 meters (one-seventh of a mile) in diameter. As one drives over the makeshift bridge into the circular paved parking area that occupies most of the middle of the ring, SPEAR looks like a series of concrete boxes loosely arranged in a circle. At the edge of the parking area is a corrugated metal building; scattered around the paved lot are some metal sheds for supplies and a trailer used for office space. Astride the ring but opposite each other are two more corrugated metal buildings. Walking inside one of these two buildings, one is confronted by a deep, paved rectangular pit surrounded by a narrow walkway with a metal railing. Sunk halfway into the pit is a 150-ton octagonally shaped object 15 feet in diameter, surrounded by people crawling around it and inside of it. This is one of the SPEAR detectors, called MARK I, later MARK II. It appears to be a huge mechanical model of an eye, with a hole for the lens and eight concentric rings of alternating grey metal and black plastic for the iris, all outlined by black metal. Masses of little wires link the rings. Radiating beyond the black metal are about 25 light grey metal rods. Between each of these are two protruding tubes. Thick wires extend out of two sockets sunk into each of those tubes. The rays and the tubes are framed by eight grey metal bars. Each bar is echoed by two behind it. These bars look like retracted eyelids, and thick electrical cables from the coalesced wires drape over the metal bars like tangled eyelashes.

What we are seeing is a series of concentric wire chambers, rings of tubular-shaped scintillation counters wrapped in a magnet, all surrounded by more counters. The hole in the center is for the ten-by-two-inch pipe that carries the two beams of electrons and positrons, which travel in opposite directions around the ring. When the beams are at the appropriate energy and density, they are deflected toward each other while they are passing through the section of beam pipe that is surrounded for 15 feet by this detector. Paths of the particles created from the energy of the electron-positron annihilation are detected, it is hoped, as they pass through these diverse media.

Retracing our steps to the parking lot, we can enter the metal building housing the Sigma 5 computer that runs SPEAR and MARK as well as collecting and analyzing data from the experiments. Watching the computer graphic display on CRT screens, one can imagine that one is seeing the tracks of particles created from the annihilation. It is rather tame compared to standing on top of a shaking bubble chamber. Nevertheless, the process being observed is less predictable. In fact, two years after SPEAR was completed, entirely new particles were discovered by the MARK I (Augustin et al., 1974). That weekend, there was a lot of champagne drunk in the control room at SPEAR, where many people crowded in to see the data reconstructed on the CRT. *Physical Review Letters* nearly refused to publish the scientific information when they learned that a student reporter had already printed comments about the discovery in the *Stanford Daily.*

These experimental results confounded existing theories and led to a Nobel Prize for the group's leader in 1976 (Bjorken, 1976). The prize was shared with another physicist at Brookhaven National Laboratory (BNL), located near New York City on Long Island. The working styles of the two groups were considered by physicists to be diametrically opposed: the SLAC group did "horseback" physics, aggressive and daring, with charismatic leadership; the BNL group was said to be "finely tuned," meticulous, and cautious, and its leadership was labeled fiercely authoritarian. Gossip was intense about exactly how the two groups could have made the same discovery at the same time. Attention focused on "leaks"; some were sure that rivals of either the SLAC or BNL leader had informed former students or research associates at the other lab where to look for the important new data. Many were startled that groups with such different styles should have produced the same results at the same time.

Many were surprised, too, at the awarding of a prize for such recent work; some physicists at the lab said that, since funding in high energy physics had declined drastically worldwide, the Nobel committee had been urged to make an award in particle physics soon, in the hopes that funding would be stimulated. In both these cases of what was considered anomalous information ("simultaneous" discovery by two disparate styles of physics and early award of a Nobel prize), physicists responded with a mass of hypotheses, just as they did to the data generated by the experiments at SPEAR. A satire on the proliferation of theories was submitted to *Physical Review Letters* by Marty Einhorn and Chris Quigg of Fermilab, naming their own theory "Pandemonium" (Einhorn and Quigg, 1975).

There were so many interpretations that *Physical Re-*

view Letters declined to publish them for a period. "Phys. Rev. Letters," as it is called, publishes brief articles on important issues rapidly, presuming that more expanded work will be published later in *Physical Review.* In 1976, the delay between submission and publication for *Physical Review Letters* was five weeks to eight months, for *Physical Review,* one to thirteen months; in 1971, the gap was more significant: five weeks to six months for the former, two to fifteen months for the later (Kirk, 1977).

NAMING MACHINES

The experimentalists announce their investment in the machines by anthropomorphizing them (e.g., the acronymic naming of machines such as LASS, PEP, DORIS, DESY, and SPEAR), but the role of the detector in making facts is not conceded. It is a common practice in the high energy physics community to use acronyms with humorous or mundane meanings in place of the longer descriptive names. Names appear to be arranged and rearranged until interesting acronyms emerge. The gap between the acronym and the name is in fact the difference between the "signifier," or pointer, and the "signified," or what is pointed to, where the acronym serves as a signifier and the apparently more accurate name serves as signified. Signifiers, appearances, are devalued as humorous and mundane; the signified, underlying regularities are regarded as essential and real.

The signifiers, the acronyms, however, are not merely humorous. As names bestowed by the group leader, they are signs of his "ownership" of the detector. For example, the name of the detector called LASS (*L*arge *A*perture *S*olenoid *S*pectrometer) always evokes the Scotsman who leads that group. In the realm of data rather than machines, the different names that two separate groups give to the same particle reveal their claims of ownership. The SPEAR group at SLAC called the particle *psi* because the highly prized computer graphics of the data produced by their detector generated a visual pattern that strongly resembled the shape of the Greek letter *psi.* Calling the particle *psi* called attention to their distinctive collection process. The Brookhaven group named the particle J because that capital letter in the Roman alphabet strongly resembles the Chinese character of that group leader's name (Ting). In one case, the name resembles nature's presumed signature on a specific detector; in the other, the name resembles the signature of a human being. The perceptual repetition of these signatures in the data continually establish their claims to its ownership. Naming data and naming machines is the prerogative of those who might be geniuses.

REPRODUCING DISCOVERY

Data from detectors will only be named a discovery if they can be reproduced at least in principle, if their production can be endlessly repeated in other detectors, given the proper coding. It is the job of the scientist to identify that coding, to show that the traces of nature in the machine are not noise, but data. Providing the correct reading of those traces will enable the making of the same traces elsewhere. In this way, the physicists prove that their reading, their physical idealization of the traces, works. Nothing else establishes the realism in these idealizations. As theorists say, "If I weren't interested in having my ideas proven real, I'd be a mathematician." Theorists need the detectors to gain that realism. The difficulty is in finding that one reading that works. The method for finding that one real reading is in knowing how to make cuts such that one is left with good data on the one hand and noise on the other, much as one learns to make cuts among novices, leaving only scientists. Knowing where to make cuts, according to the experimentalists, means knowing one's detector and the only way to know a detector is to build it.

If the cuts are made properly and the remaining data are significant and reproducible in other detectors, especially other kinds of detectors, then the data constitute discovery. The meaning of the word "reproducible" here is problematic: no two detectors are alike. No one could get funding to build a copy of another detector and no one would want to try. There would be no credit and influence to be gained. Furthermore, only the group that built the original would have the knowledge to build the copy.

There are different models of how to design a detector. One of the detectors at SLAC, LASS, was designed to enable meticulous analysis. It is an elegant set of equipment, presumably collecting very clear data. That group's leader said that it was necessary to see a detector, not as an end in itself, but as a piece of equipment designed to "resolve a philosophic question." He sees the strength of his group and its detector as the power to correlate data analysis with physics questions. Collecting clear data is crucial to that goal of resolving questions; the group is committed to finding data without noise.

The group at SPEAR sees itself as having designed its detector architecturally rather than analytically. The group regards its detector and their physics interests as evolving. The detector is designed so that it can be quickly altered. "We're too stupid to build it right from the beginning, but we can build it so that it can be fired easily." This physicist means that their group was smart enough to build ways of correcting their "error bars" into the machine itself, refining, deconstructing, and reconstructing the machine as data analysis improved: "If the detector's architecture is good, new parts will fit in."

He contrasted this floor approach to a third group's detector: "Our detector [SPEAR] was built on much less money, and we are better for it: we built with much more thought and ingeniuty. Their machine [End Station A] was built in fat times, and you can still see it in their cupboards. If you wanted three of something, the leader said to order a hundred; we will use them eventually." The implication was that the third group was just mechanically replacing its detector because it had so many spare parts. The third group made no claim to elegant equipment or subtle architecture. It had built its reputation on quickly getting finely structured results to a specific set of questions. Their detector's results are widely regarded as reliable because the machines are considered to be overbuilt.

The difference among these detectors serves as a mnemonic device for thinking about the difference between the various groups' models for how to elicit traces from nature that are both significant and reproducible. Detectors themselves, then, are a system for classifying modes of discovery. Each is the material embodiment of a research group's version of how to produce and reproduce fine physics, gaining a place for the groups' work in the taxonomy of established knowledge in physics. But each of the groups' strategies for finding traces is in fact a strategy for dealing with noise. That is, the first one is spare and elegant, meant for philosophical questions. The second is an ingenious architecture, meant for reconstruction and deconstruction. The third is fat and overbuilt, meant to be reliable.

REPRODUCING SCIENTISTS: THE PREDICTABLE READER, THE IDEAL READER, AND THE BRICOLEUR[6]

Detectors tell us not only about the production and reproduction of discovery but also about the production and reproduction of physicists. In learning to do an experiment, novices learn to make good physics and to be good physicists. In other words, the patterning of culturally desired characteristics of individuals and relationships occurs in the context of doing an experiment, much of it by means of non-verbal, kinesic discourse. For the sake of the stability of a community, information about its human relationships and its sense of the nature of human society should be imprinted on its novices in a way that is least subject to loss or revision (Bateson, 1972b; see also Bateson, 1979). Non-verbal discourse is often unconscious, and hence ideally suited to messages of socialization about relationships.

In other words, the diurnal practice of doing a physics experiment not only structures the information taught to the novices but also, as Gregory Bateson says, "structures the recipient of the message" (Bateson, 1972a). Clifford Geertz makes the same point: "The ability to respond . . . is no less a cultural artifact than the objects and devices concocted to 'affect' it. . . . [These capacities] are brought into actual existence by the experience of living in the midst of certain sorts of things to look at, listen to, handle, think about, cope with, and react to" (Geertz, 1983). Within modern literary theory this same phenomenon is discussed in terms of how fictive and non-fictive texts construct, imply, or prefigure their readers' responses.[7]

Familiar environments have a compelling power to instruct us, kinesically, according to Gaston Bachelard:

> It is a group of habits. . . . The feel of the tiniest latch has remained in our hands. . . . But we are very surprised, when we return to the old house, after an odyssey of many years, to find that the most delicate gestures suddenly come alive, are still faultless. In short, the house we were born in has engraved within us the hierarchy of the various functions of inhabiting that particular house, and all the other houses are but variations on a fundamental theme. The word habit is too worn a word to express this passionate liaison of our bodies, which do not forget, with an unforgettable house. (Bachelard, 1969)

I have argued elsewhere that the buildings, the landscape, and the views of the lab are parts of an intricate web of associations for the physicists who know the laboratory as well as their homes, reminding them, inextricably, as they move, of who they are and how they must act. I am also claiming that Bachelard's house is like the fat and reliable third detector, like a work with one proper reading, and that work constructs its readers authoritatively. Nobel prize-winning work was done at this machine too, and other physicists say of the Nobelist that he merely had "his thumb on the button at the right time." The machine makes the scientist, and that kind of scientist makes that kind of machine.

Now, contrast Bachelard's walks with the observations of Yi-Fu Tuan, the eminent social geographer, on the view of nature imbedded in the architecture and settings of country villas in the Renaissnce:

> Posed against the city are nature and the countryside. . . . This nature is viewed essentially in two ways: aesthetically, as the setting for a country villa, for quiet study and exhalted philosophizing; and morally, as the stage for the development of independent and manly virtues. (Both attitudes overlook the problem of livelihood, and condescend not only toward the money-loving herd of the cities, but toward the pig-and-chicken farmers and the confined worlds of small towns). The aesthetic view of nature is characteristic of the upper class of both Western and Oriental societies. (Tuan, 1970)

Like these aesthetically pleasing landscapes the spare and elegant second detector is designed for contemplation, for the construction of aesthetically articulate viewers of nature, perfect visual texts generating cultured, ideal readers, or viewers, of nature. This group has not discovered; it has refined.

The literary critic Umberto Eco has argued that certain kinds of texts neither determine the reader's response nor assume ideal, aesthetically informed readers. These texts induce the reader to take "inferential walks":

> At the level of discursive structure the reader is invited to fill up the various empty phrastic spaces (texts are lazy machineries that ask someone to do a part of their job). At the level of narrative structures the reader is supposed to make forecasts concerning the future of the fabula. To do this the reader is supposed to resort to various intertextual frames among which to take his inferential walks. Every text, even though not specifically narrative is in some way making the addressee expect (and foresee) the fulfillment of every unaccomplished sentence. (Eco, 1979).[8]

Eco's words recall those of the physicists who wanted to build "architectural" detectors responsive to rearrangement, to deconstruction and reconstruction. The scientists who labor at those detectors see themselves as underdogs, witty, wily shrewd at their craft. The Nobel prize-winning work done at this machine has made heros of the scientists there. They are busy concocting new schemes, other "Rube Goldberg" machines. They are playful; they are *bricoleurs* (dodgers, shufflers, jacks-of-all-trades).

I have just argued that detectors can be a sign of difference among physicists, a sign of ownership, a sign of discovering strategies. However, the detector is also an emblem of objectivity. It is the detector as a transparent recording device, passively recording the signature of nature, much as a camera is popularly thought to have the capacity to represent the world as it really is, that establishes the legitimacy of data as facts.[9] This "hard" technology, then, is produced by scientists to authorize the construction of scientific facts; this aspect of the means of their construction is then obscured and excluded in the analysis of those facts.[10] An ideal detector that has gained the status of producing knowledge ideally is remembered as a transparent scientific instrument that passively and objectively records information about and from nature. Nature is thought to have inscribed its signature on those specifically designed detectors. The data in the machines are thus "naturalized"; the data *are* nature.

Like many high energy physicists, Descartes claimed a correspondence between mind and nature; he presumed, as they do, that reason and order are inherent in both mind and nature. For him, the media for acquiring information about nature were the senses (which were recognized as faulty). The guarantee that what mind apprehends from this transient sense data actually corresponded to nature was God. But Galileo and Bacon believed scientific instruments could assist God in assuring the validity of one's data. Presumably, machines could reproduce data that would resemble that of the ephemeral senses, but without the "subjective bias" of biological senses. I am arguing that, *just as God* was the mechanism by which Descartes assured himself of the veracity of his subjective sensations and observations, *the machine now* affirms the validity of the physical idealizations proposed by the high energy physicists by denying the role of human agency in the construction of scientific facts.[11]

It is theorists who are more likely to see detectors as scientific instruments that simply record nature, as transcription devices that themselves leave no trace. These rather Platonic theorists, unlike the Cartesian experimentalists, see the data produced by the detector as being uncontaminated by the machine, if they have been assured by the experimentalists that the experiment was properly designed and conducted. Theorists themselves have little access to, or knowledge about, the detectors. The papers given by experimentalists at seminars and conferences begin with a detailed description of their detector, and devote at least a third of their presentations to these machines before introducing the data generated in their experiments and reporting how those data were analyzed in order to produce "curves" (interpretations that have an acceptable degree of "fit" with the data). Nevertheless, the theorists rarely attend carefully to the first part of the talk, to these "technical details," referring to them as the "Scotch tape" part of the talk.

For the experimentalists, the detector never ceases to be very expensive congealed human labor. It is the product of the group, its representation, and its signature, just as the data in the detector is presumed to be nature's representation and signature.[12] Experimentalists read their detectors not only as recording nature but also as mnemonic devices for the past, present, and future of the research group. The detector is the visible sign of their scientific genius. However, a detector can be the reassuring sign of objectivity only when as a machine it is invisible.[13]

OBJECTIVITY AND TRACES OF REASON

In this ethnography, my attention has been drawn away from physics, as an "object" of knowledge, toward the physicist, the "subjects" of knowledge. Science is a world in which all questions are ideally reducible to problems of one variable (by means of a set of conventions whereby certain aspects of an event can be excluded from what needs explaining). I shall argue that it is in this exclusion or disavowal that we can learn how scientists come to see themselves as particularly suited to comprehend nature.

They see each human being as a composite of rational and irrational elements, a combination of order and disorder. They also believe that the ratio of rationality to irrationality is randomly distributed among humans. They also believe that the shape of any random distribution is determined, and hence knowable. According to these laws of random distribution, they feel that there would always be a small proportion of humanity with an exceptionally high ratio of rationality to irrationality and that such highly unusual people are uniquely suited to become scientists. (They understand that these unusual people may not all become scientists, but all scientists come from this "gene pool.") From this series of definitions, we see that scientists are not made; they are born, revealed as scientists in a series of exclusions made in the course of scientific training.

Nature, like humans, presents the appearance of continuous flow, but underlying the appearance of disorder are regularities in the form of immutable mathematical laws. The process of reduction (the classification of all phenomena according to certain attributes, such as rationality and irrationality, cause and effect, and excluding all others) is seen as a necessary first step in achieving an understanding of this inherent order. That which eludes reduction to the chosen attributes, whether in nature or in scientists, is avoided, rejected, or considered merely amusing. Second-order regularity is necessary for nature to be knowable and for scientists to be identifiable. "Progress" in science and in the education of scientists consists of unmasking these regularities, not in constructing them.

In nature the same forces exist and the same elements exist in the same proportion for all time. Both the periodic table of the elements and SU3, the classification system of fundamental particles, display immutable elements and their relations. All apparent flux is merely the redistribution of basic elements; the fundamental particles appear to be transformed into other particles, but these transitions, or "decay modes," are only patterns of possible recombinations. The scientific identification of cause and effect would suggest an inherent temporality; however, since the effect has been defined as inherent in the cause—just as a particle's decay modes are inherent in its definition—all is determined. Like scientists, nature awaits discovery: nature is not becoming orderly, it is orderly; scientists are not becoming more rational, they are rational.

By contrast, whatever is inadequate, inappropriate, or imperfect will and should perish. In particular interaction, instabilities are short-lived and rapidly move toward stable configurations. Much incorrect work is done by scientists, and may even be accepted for a time, but this is ephereral.

The language of physicists is rich in metaphors for change that have negative connotations, and there are abundant metaphors for stability that have positive connotations. The key metaphors refer to the "simple initial state" that must be achieved in any good experiment, albeit with great difficulty. In a particle physics experiment, this desired "ground state" first requires meticulous detector design and construction, a process that takes at least months, and often years. In addition, there must be precise computer analysis of the statistical errors and systematic errors that are inherent in the design and operation of the accelerator, the detectors, and/or the experiment. Then the group must learn to "run" the experiment in a painstaking and exact manner, constantly calibrating the accelerator beam, the detector itself, and the data collection procedures.

In this way, the experimentalists decide what part of their data can be considered valid, and what part must be "cut," ignored as "noise." It is all of this activity taken together that defines the "ground state." Beyond the generation of this "simple initial state" comes the analysis of data collection by experiment. Interpretation of these data is quick and straightforward only in rare cases. Usually, this process takes a year or more, and even then the results are scrutinized carefully by the community and accepted only when confirmed by data from other experiments. It takes human, linear time to pull the physics out of the morass of data. For other experimentalists to be interested in corroborating the data from another experiment, they must be convinced of the significance of the results. This is accomplished by arguing that a precisely defined "ground state" was generated in the experiment.

In the particle physics community, this model for making fine physics is also the model for making good physicists, and for making a proper environment for physics. In producing postdocs, running a laboratory, or designing experiments, "predictable, smooth behavior" must be generated before instabilities can be regarded as innovative. It is only in the context of an elaborately defined "ground state" that exceptional ideas, institutional change, or data fluctations can be labeled original, innovative, or significant.

It is his concern for this "coherent ground state" that leads a senior physicist to demand that postdocs be meticulous and thorough, hard and willing workers who have the patience to do a very long and tedious job carefully. This concern also provides justification for directors of laboratories who resist all change in their organizations. In spite of their efforts to maintain stability and establish truth, uncertainty and error remain a part of scientific work. This is the consequence, according to the scientist's "world view," of a fatal flaw in scientists: as human beings, they reflect some admixture of irrationality, even if in minor amounts compared to the rest of the population. That is, pure objectivity is tacitly recognized as impossible. The degree of error, however, can be calculated, and this error can be minimized. The means is peer review, or collective surveillance: the final degree of order, oddly, comes from human institutions.

Scientific endeavor (striving for objectivity in the study of nature) requires resources from the rest of the society in the form of student funds. These resources are procured through activities that are potentially contaminating (teaching, administration, and consulting for the government) because they require the cultivation of skills not thought to be based on reason—in particular, the power of persuasion. As a consequence, these activities are thought to be done properly only by senior scientists who have demonstrated their resistance to enticement of such pollution.

Nevertheless, this contact with irrationality renders these people incapable of creative work in science. The scientists themselves usually claim that they no longer do research because they have no time; other scientists believe that these science-statesmen chose their new role because they no longer had "any science left in them," a loss of rational potency. These statesmen regularly transgress the boundary between the domain of rational laws of nature and the domain of arbitrary laws of humanity, but not with impugnity. Becoming an emissary to the world of the merely human, they are disbarred from practicing science. In a final twist of irony, they take their place in the margins of the textbooks of undergraduate physics students, heroically guarding the boundaries of physics.

I have argued that scientists come to see themselves as sharing with nature a certain crucial quality. They regard nature, underlying the appearance of continuous flux, as being a system of regularities patterned according to immutable laws that can be articulated in mathematical form. In the same way, scientists are seen as human beings who are exceptionally rational, with a minimal (if any) admixture of irrationality. The regularities are not constructed in nature, but discovered; the rationality in scientists is not created, but unmasked in a training process that progressively and systematically excludes the irrational.

The students learn this process of exclusion in a series of stages, beginning with what is excluded from the main body of their textbook and set into the margins. By learning to read the margins, the novices learn to become competent practitioners of the culture of the particle physics community. Graduate students are learning the boundaries of their community and, as they say, the "fine structure" of the differences within it, and postdocs learn how to comport themselves in situations that reflect these differences. Because they are at major labs, they are in situations that reflect these differences. They are learning that outsiders (experimentalists) are devalued and exactly how this is done and what justifications are given. Work unlike that done at the major centers of research is easily labeled peripheral; difference within the community becomes redefined as eccentricity or deviance. That is, the novices are learning to distinguish outsiders from insiders. They learn in detail the varieties of being an outsider that are available to them as career choices. The postdocs are learning to be anxious about the consequences of their evaluation and to believe in the justice of that evaluation. Hence, one learns the fine structure of hierarchy before one learns one's final place in it.

The margins stand as boundary markers, defining the community by defining specifically both what the group wishes to include and exclude. In the analytic language developed by Gregory Bateson the margins serve as "context markers," operating at a "higher logical type" than the (social) text, indicating to those who know the cultural clues in the context markers how to read the messages in the (social) text (Bateson, 1972c). In other words, the immortal heroes of science in the margins of the undergraduate physics textbook are "context markers" defining the posture one must display and the genealogy one must acquire in order to become a producer of (machine) texts, in time.

To become a designer of machines one must survive the exclusionary cuts, show oneself to be part of the solution, not part of the noise. In their quest for survival novices learn about time. During the undergraduate years, students discover the insignificance of the past in high energy physics. Knowing the formal history of their field (especially the record of "erroneous" ideas) is considered debilitating. The past is thought to be devoid of any significant information about the current canon of physics. Students also discover that there is a great gap between the heroes of science and their own limited capacities. Their study of the history of science is condensed into the celebration of timeless genius and reproducing the successful experiments.

Graduate students learn to fear "losing" data (e.g., accidental errasure of computer records, failure of detectors during an experiment that results in loss of beamtime) and to fear not having enough time to incorporate enough information to do their work well. These students learn stories about other novices who have heroically conquered these threats by getting, recording, and saving data when time is running out. Postdoctoral research associates begin to learn and communicate about physics orally, rather than through books and articles. They hear stories about those who have "made it"; they realize how important it is to anticipate the future new directions in theory and new solutions to the design problems of detectors. In order to be successful, they must figure out the future of physics.

When the novices become members of a research group, they begin to identify their own careers with that of a detector, if they are experimentalists, or with a set of models (e.g., field or S-matrix theory), if theorists. After their ten years of training, they have about ten more years to make their reputations in the field. It is as new group members that the young physicists learn of the significance of the lifetimes of detectors, research groups, laboratories, careers, and ideas.

Fear of obsolence in these five areas leads to a recognition that uptime/downtime and beamtime are commodities to be acquired and used for power in one's contest with obsolescence. During the first three stages of training, however, the novices learned to deny the long-term significance of human time. Consequently, all interest is focused on getting beamtime and discovering the immutable laws of nature, privileging an eternal present in which loss of the past, loss of data, anxiety about the future, and fear of absolescence are repressed.

The contrast between soldiers in the field and timeless genius, between mortality and eternal principles, could not be more stark. On that thin line between anonymity and immortality stands the joker and broker who may be a genius, the group leader who barters time.

LABORATORY TIME

A group leader tries to barter six kinds of time: (1) "up" and "down" time, referring to whether or not the accelerator beam is running; (2) beamtime, which refers to scheduled access to some portion of the beam; (3) lifetime of a detector, from gestation to obsolescence; (4) lifetime of a laboratory, from design and funding to the phasing out of particle physics research; (5) career time, meaning the proper amount of time one should spend in each level or position, and also the period of time one can be expected to be a productive physicist; and (6) even the lifetime of an idea. All six units of time organize the laboratory, and power in the laboratory is based upon these experiences of time. The six units of time can be reduced to two general types: replicable time, which can be accumulated, and calendrical time, which slowly slips away.

The daily rounds of the experimentalists are very different in the two phases of the calendar known as uptime and downtime, rather like the differences in Nuer social organization in their two alternating ecological settings (Evans-Pritchard, 1940). While the accelerator is up and a group's experiment is running, the experimentalists spend most of their time in the research yard near their detector, monitoring the course of their experiments, including detector performance and data collection devices. The detectors are notoriously temperamental; if the detector malfunctions during a run, the group's scheduled allotment of beamtime will be officially forfeited until the detector can be repaired. (The group's beamtime may be returned later through informal arrangements if another group's detector malfunctions.) Beamtime is a term that in fact refers to the number of bunches per second to which a group has access. Beamtime and uptime/downtime are calculated in hours. At SLAC, these hours are a resource to be spent sometime during a run—preferably according to a prearranged schedule. Finally, new group members learn how to get beamtime, and to regard time as a commodity, a nexus of social relations, that is negotiable and can be exchanged.

Five other social units of time in the research process

are measured in years, and referred to as "lifetimes." (This term is an allusion to the "decay times" of particles, which is a measurement of the characteristic lifetime of a particle.) In general, these lifetimes are slowly "running out." The total lifetime of a detector from gestation to obsolescene is approximately twenty years. Any given configuration of detector lasts two to five years; it is modified as the group develops new physical, analytical, and mechanical techniques for their detector. The property of a research group is its detector and its intimate knowledge of that detector. Most groups cohere as long as their detector is considered to be "producing good physics" economically, in terms of money, beamtime, and career time. Hence, the lifetime of a detector is strongly correlated to the productive lifetime of a research group. (Once a detector becomes obsolete for particle physicsts, it may yet have a long, useful life as a research device in other fields. In fact, that sort of "spinoff" is one of the ways a laboratory justifies its budget to national funding agencies.)

In the United States, research groups do not cohere across generations; hence, they do not outlive their detector. A group's wealth is embodied in its detector (from which it derives a network). The lifetime of an American research group is measured calendrically and hence is emphemeral; the anxiety about time is translated into tension and competition *within* one's own generation among groups for beamtime.

The lifetime of a laboratory marks the time from its conception to the end of particle physics research and commencement of applied physics research. This lifetime is on the order of twenty years. A laboratory may extend its lifetime by undertaking major modifications of its accelerator (to increase its peak energy and refine other beam parameters) and replacing its attendant detectors. SLAC is an example of a laboratory attempting reincarnation, through the construction of PEP (Proton-Electron-

Positron). Argonne National Laboratory in Illinois and Lawrence Berkeley Laboratory in California are examples of facilities that now do a wide variety of research with little "state-of-the-art" high energy physics.

There is also a strong sense of the proper amount of time one should spend in each position during one's career as a physicist. For example, three to five years as a postdoctoral research associate is suitable; ten years, although feasible, is not suitable. Physicists prefer to spend the shortest time possible in each of the stages up to research group leader. This is because they assume that there is a rather short time (about twenty years) one can expect to be a productive physicist; after this period, many physicists increase their teaching and administrative activities. The notion of the productive lifetime of a physicist is replicated in the measurement of the lifetime of an idea. If a theory (or datum) proves useful, it becomes a tool for identifying new theories (or data). If the theory (or datum) does not prove useful rather quickly, it is dropped.

These social units of time in high energy physics (uptime/downtime, beamtime; lifetimes of detectors, research groups, laboratories, ideas; and productive career time of a physicist) can be reduced to two kinds; the first recurs, and the other is ephemeral. An example of the first—at least in the American labs—is beamtime. Access to beamtime means access to space in the research yard; similarly, access to detectors generally means access to beamtime. This beamtime is a negotiable commodity, and is a source of power. It is measured in seconds, hours, days, and months. For the Americans, these allotments can be accumulated and exchanged. In the other construction, time is calendrical, serving as a parameter marking a nearly irreversible sequence of events (such as the decay times of detectors, ideas, and physicists). These times are measured in years, and are irreplaceable. Between replicable and ephemeral time stand resurrected, modified machines, ideas that become discoveries, and physists who were named geniuses.

ORDERING TIME

Particle physicists share a set of meanings about time. These conceptual assumptions are examples of that aspect of Kuhnian paradigms called "symbolic generalizations." Diana Crane has identified such generalizations as the "body of knowledge (laws of nature) which is accepted by all members of the field and which has not been questioned for decades" (Crane, 1980).[14] Those meanings extend from their physical theories about time to the ways time orders the activity of the physicists. In each of these domains, there are two apparently contradictory notions of what time is. Detectors display the tension between these different kinds of time; they also provide a model for the resolution of those tensions.

Very briefly, high energy physicists agree among themselves that there are probably only two kinds of time: time in a non-relatvistic setting, and time in a relativistic setting. Time in a non-relativistic setting is simply a marker. It is only a "milestone" marking a sequence of events in space. This space can be any designated, arbitrary spatial reference frame, including a three-dimensional world in which humans act. For example, the experiments conducted by scientists in the laboratory occur in a non-relativistic setting; that is, time functions as a marker. This kinematic description of the world applies when the speed of motion is very small relative to the speed of light.

In a relativistic setting, time becomes another coordinate, interchangeable with space. Time is elevated to a status equal to that of space instead of being simply a milestone helping us to find our way through the environment. This non-relativistic time thus serves as a limiting case for relativistic time, which applies in systems near the speed of light. Thus, the realms in which each definition prevails can be differentiated, but they are, nonetheless, contiguous.

For example, in an experiment, human signals to the

research apparatus occur in a non-relativistic frame, events occurring in the apparatus are in a relativistic frame. Hence, the experiments are being conducted in both frames "simultaneously." It is essential for the physicists to accept this conjunction in order to get on with the research at hand. (Both concepts of time are taught in first-year physics classes.) The research apparatus, the detector, is the locus of this conjunction of the physical theories of time.

CHARTING KNOWLEDGE

The physicists' charting of particles and their relations is modeled on Mendelev's periodic table of the elements in chemistry. The characteristics used to define the particles are their lifetimes before they decay into the other particles, their massed, electrical charge, and their means of interacting with other particles (via the strong, weak, magnetic, and gravitational forces). These qualities are determmined in the detectors by measuring tracks of particles for their radius of curvature, their time of flight, and their interaction with other particles. These basic constituents of matter are revealed by spatial display of their qualities in detectors, an array of "signs in their spatial simultaneity" (Foucault, 1970). The taxonomic system of knowledge in physics annihilates the past and future in an eternal present, an infinite repetition of a limited number of physical processes.

The other orientation to time in the high energy physics corpus of knowledge is the resurrection of time in the presumably progressive accumulation of knowledge about that taxonomy. In the idea of progress entertained by these physicists, science is the prime agent. Scientists conduct explorations of nature (which is seen as a hitherto unknown, untamed wilderness) in order to produce scientific ideas that will be used by society to control and domesticate nature.[15] It is through predictions about future events that scientists believe that their ideas are confirmed. These validated theories then represent progress over the past's lack of knowledge. Thus, this model of progress is a *post hoc* reordering scheme.

On that thin border between progress and taxonomy is the new discovery. Or, as the physicists put it, poignantly, "Last year's discovery is this year's calibration." Any discovery will be inserted into that accumulating body of physical knowledge, their taxonomy of nature. The greatest discoveries are those that alter the taxonomy, creating new gaps in the classification system. When this happens, one will have many imitators, scientists who will be "looking for all the other animals in your zoo." If the data are real, they will fall within the classification scheme, the taxonomy of the predictable, the repeatable. I am asserting that Foucault's definition of the classical *episteme* is an apt description of the classificatory system of knowledge that prevails in high energy physics: "The Classical *epsiteme* . . . [is] an exhaustive ordering of the world . . . directed towards the discovery of simple elements and their progressive combination; and at their centre they form a table on which knowledge is displayed in a system contemporary with itself" (Foucault, 1970). The present has become a thin border between past and future, between an as yet untamed and an already domesticated nature. (Levi-Strauss, 1966, p. 220 and passim).[16] In their firm commitment to the inevitability of progress of their field, physicists appear to be valorizing an ephemeral model of success, generating obsolescence. Nevertheless, invariance, stability, and symmetry are more highly valued in this field. The physicists' search for these constants appears to me to be a quest for something ahistorical, acultural, and non-contingent on which to base all human understanding. History, culture, and subjective experience are seen as so many flaws that ultimately can be eluded by means of science. It seems to represent a desire to stand outside time, apart from the irregularities of experience and appearances. The belief in prediction becomes pleasure in control of the future, the faith in progress, a desire for domination of the past. The past, the future, and the setting in which humans act and produce events are devalued. The mere sequence of events in this non-relativistic frame is simply a parameter that marks change in some arbitrary—and, hence, insignificant—spatial frame of reference. This "non-relativistic" time of human action is conquered repeatedly by the "relativistic" frame of scientific knowledge. I am not saying that cosmology determines the social organization of time or vice versa. I am saying that both domains are similarly patterned.

I suggest that this Platonic rejection of flux corresponds to a nearly electrifying tension about time that is coiled at the center of the high energy physics culture. That anxiety about time is an avoidance of the insignificant past, the quickly disappearing present, and a too rapidly advancing future, and a fear of obsolescence.

THE CONVERGENCE OF TIME IN DETECTORS

The detector serves to reconcile a paradox between immutable, repeating time and ephemeral, linear time. This reconciliation has been achieved by establishing a certain kind of relationship between apparently contradictory beliefs and experiments, between time templates and time anxieties (see Figure 2-1). In the domain of *cosmology* the tension is between relativistic and non-relativistic time. In the domain of *knowledge,* there is a conflict between a taxonomic system of knowledge, where the laws of nature are not believed to change, and a belief in the progressive accumulation of that knowledge. Third, nature, which is seen as supremely uninfluenced by the opinions of human beings about how it operates, is best understood by a special group of humans called *scientists.* Finally, in the *laboratories,* immutable, repeating time, like beamtime, can be accumulated and bartered, but ephemeral, linear time, like lifetimes, only can be lost, irretrievably.

	Stabilities	*Instabilities*	*Practices*
Cosmology	Relativity	Non-relativity	
Knowledge	Taxonomy	Progress	This year's ideas are next year's calibrations
Scientists	Timeless genius	Soldiers in the field	Group leader
	Ideal readers	Predictable readers	*Bricoleurs*
Laboratories	Replicable time	Ephemeral time	Modifiable machines
Detectors	Philosophy *Data*	Engineering *Noise*	Architecture

Figure 2-1
Physics Community Constructions of Time

Detectors are at the juncture of each of these paradoxes, simultaneously articulating and resolving the contradictions. Good detectors are, by their definition, imperfect devices for receiving signals from nature; nevertheless, detectors (forgotten and invisible) come to be seen as transparent scientific instruments for reading news from nature. But detectors are also mnemonic devices for how to make a discovery. And they also signal the past, present, and future of a research group. The detectors are very expensive congealed human labor, a product of the massive organization of human resources and capital. The detector does not determine the organization of scientific research; it does not determine what is discovered; it does not produce scientists; and it does not determine the organization of knowledge. However, detectors are at the nexus of all those activities, patterning organization, scientists, and discovery, patterning contradiction and patterning the resolution of contradiction.

That pattern can vary, as I discussed when I identified

three styles in the design of detectors as different strategies for making discoveries, making scientists, making experiments, and representing nature: the elegant LASS, shaped for contemplative reflection; the overbuilt End Station A, coding reflexive readers; and the architectural PEP, designed for play.

However, the symbol—the detector—that reconciles these contradictions itself ideally becomes invisible and is forgotten. It is the gap at the core of the community. That is, the role of human action in the construction of scientific facts is the detector. It is at the site of the detector that the presence of human action is forgotten. Heinz Pagels said that "nature is the text, a cipher to be deciphered" (Pagels, 1983). To scientists, especially retrospectively, their machines are transparent looking glasses, merely spectacles for regarding the immutable fixed text of nature. The physicists want to see the data and nature as equivalent. The experimentalists want to see themselves as the ghost writers of that story about nature, but not the authors. The theorists want to see nature as the author of the data text. To Latour and Woolgar, describing immunology, the text is the data and the machine is its authoritative author; for them scientists are astute decoders of data texts (Latour and Woolgar, 1979). In my view machines are the texts, variable texts that enable us to watch the reproduction of discovery, the reproduction of nature, and the reproduction of scientists, as scientists engage themselves in the incessant production and reading of machines in which neither the text nor the reading is fixed. In other words, there is no stable text and there is no fixed reading; there are many kinds of discovery modeled in detectors. Instead, the texts and reading situated in the practices of high energy physicists all address the production of discovery, the production of nature, and the production of scientists. These texts and these readings respond to and generate traditions about how to identify discovery, interpret nature, and evaluate scientists. Reading detectors has enabled us to produce a reading of the high energy physics community's culture.

One reader of this paper suggested that these physicists could not be as unaware of the "philosophical" implications of their relativistic theories and experiments as my representation of them implies. I had the same expectation. When I asked physicists about this, however, they all told me that they were "completely uninterested in philosophy." Actually, many expressed themselves much more vehemently. Their scorn for discussions of relativity by non-practicing high energy physicists was always palpable; they were also quite easily bored by any effort by one of their peers to discuss relativity outside the realm of immediate theoretical or experimental issues. I found that those very few who were interested were almost always regarded by their colleagues as "odd" or "not real players in the game"—what sociologists call marginal types.

Furthermore, some readers have thought that I have represented these scientists as producing (merely) fictions. I have claimed that they produce machines that make data and that they construct narratives about their machines and their data that they take to be realistic representations of the part of our world we call nature. I am not in a position to evaluate the truth claims of the physicists' representations. I am in a position to analyze the temporal organization of their enterprise and the ethos that justifies it. I will add that I have never met a high energy physicist who did not believe that electrons "exist," unlike some philosophers. I can sympathize with the physicists, because I do believe that physicists "exist," unlike some of my more reflexivist colleagues. Nevertheless, unlike the physicists, I do accept the intellectual significance of the question, Where does the social category of "physicist" or "physics community" exist? In the final analysis the truth status of my interpretation, like that of the physicists, is open for discussion, as are the definitions of "truth" and "existence." Again, it is my experience that almost all high energy physicists, as well as MIT students, find such questions very tiresome.

I do wish to remind the reader that it is the first job of an ethnographer to engage in a willing suspension of belief (in contrast to the proverbial theater-goer who engages in a "willing suspension of disbelief"). What might be disturbing to any reader of this chapter (or others in this volume) is just that ethnographic suspension of belief in the subjects' commonsense world. We might accept that suspension easily when the anthropologist is describing some so-called "primitive" people in an "exotic" setting or some "backward" culture, remote from our own world view. I suggest that our facile acceptance of the ethnographer's suspension of belief in the world view of her subjects in those cases can conceal simple, perhaps unconscious, condescension. When the ethnographer brings this same analytic distance to the study of our own post-industrial society, it can be very disturbing. The "common sense" of our world is buttressed by a profound belief that it is "scientific," meaning "true." Or, as the physicists whose sense of certainty about this world view I have been studying would say, "it must be true because it works." This assumed correlation between our efficacy in the world and our explanations of that efficacy is not the subject of this study. To examine the common sense of time among scientists, however, is to examine the structure of our own common sense and to examine our own analytic assumptions in the study of social action.

It is a truism that the sense of time in industrial (indeed post-industrial) societies distinguishes us from our fellow human beings living in agricultural or food-gathering or hunting societies. If so, it is time, so to say, that we began to study what industrialized people make of time in their daily rounds, and to study those people who define time for us most authoritatively: scientists and engineers. I will conclude with a physics graffiti about time, space, and authority. "The speed of light is 186,000 miles per second. That's not just a good idea; its the law!" Defining time is defining law and order. Or as another wit put it, "Time is nature's way of keeping everything from happening all at once."

NOTES

1. The title of this chapter and its inspiration come from Benjamin (1969), pp. 217–251.

2. The social construction of time and space is, by now, a conventional topic in anthropological ethnographies, the most notable of which is Evans-Pritchard's study, *The Nuer* (1940). More recent studies include Bohannan (1967), pp. 315–330; Leach (1966); Pocock (1967), pp. 303–314; and Thornton (1980). See also the review article on this subject by Jack Goody (1968), pp. 30–42. Informative works by non-anthropologists include Berger and Luckmann (1966); Marcuse (1956), p. 231; Thompson (1967), pp. 56–97; Schutz (1975); *History and the Concept of Time* (1966); Lang (1975), pp. 263–280; and Meyerhoff (1968).

3. For an important discussion of the place of representation, reality, and appearance in science as well as the significance of the production of experiment, as opposed to theory, see Hacking (1983).

4. In my study of style in detectors I am indebted to the analytic, anthropological work of the archeologist Heather Lechtman. See especially Lechtman (1977, 1984).

Walter Vincenti has made a major contribution to my understanding of the relation between tacit, prescriptive, and descriptive knowledge in science and technology in his analysis of innovation, design, and production in engineering (Vincenti, 1982). On the ethnographic observation and analysis of engineering practices I have been influenced strongly by L. L. Bucciarelli of MIT. See Chapter 3.

5. See Geertz (1973b), pp. 3–30.

6. In *The Savage Mind* (1966) Claude Levi-Strauss introduces the term *bricoleur* (jack-of-all-trades, handyman, Rube Goldberg) as a definition of the activities of people in nonscientific societies who patch together explanations in an *ad hoc* fashion, unmindful of any need for articulated, abstract, analytic criteria for argument and evidence.

7. For a survey of these ideas, see Iser (1978), especially pp. 20–50; Suleiman and Grosman (1980); and Tompkins (1980). The latter two books contain extensive, annotated bibliographies.

8. Paul Ricoeur has claimed in a paper that has had considerable influence in anthropology (1971, pp. 529–562) that social action can appropriately be analyzed as a text. This theme has been developed in anthropology primarily by Clifford Geertz (1973a, pp. 448–453, 419). In cognitive anthropology a similar concept, that of competent speakers, which was adapted from Chomskian linguistics, has been used to define how actors acquire and display competence in social situations (see Frake, 1976, pp. 87–94).

9. For an analysis of "inscription" and "materialization" in science, see Latour and Woolgar (1979), pp. 245, 240. On the social construction of technology, see Mulkay (1979), pp. 70, 77.

10. See Latour and Woolgar (1979), p. 245.

11. On the denial of human action and the demand for mechanization in the production of scientific facts, see Zenzen and Restivo (1981), p. 19. See also Giedion (1969).

12. On this point, see Althusser (1971), pp. 162–166.

13. For an explication of the relation of the signified and their signifiers, see Barthes (1977).

14. For a discussion of these two constructions of time, see French (1968); Sklar (1977); and Feynman (1967). For those who prefer an introduction to these ideas at a very elementary level, see Haber (1956); for those who prefer to be amused, Winsor (1958). I have consulted with Henry Don Issac Abarbanel, theoretical physicist, Scripps Institute of Oceanography, in constructing my exegesis on these physical theories of time. Errors, of course, remain my own.

15. On domesticating a nature that is represented as female, see Keller (1980), pp. 299–308. See also Leiss (1974).

16. See Clifford Geertz (1973a) for a criticism of Levi-

Strauss's work as being itself sychronic and taxonomic. See also Ellen and Reason (1979); Goody (1977); and Needham (1979).

REFERENCES

Althusser, L. 1971. *Lenin and Philosophy.* New York: Monthly Review Press.

Augustin, J.-E., et al. 1974. "Discovery of a Narrow Resonance in e + e-Annihilation." *Physical Review Letters* 23 (Dec. 2).

Bachelard, Gaston. 1969. *The Poetics of Space,* trans. Maria Jolas. Boston: Beacon Press.

Barthes, Roland. 1977. *Elements of Semiology,* trans. Annette Lavers and Colin Smith. New York: Hill and Wang.

Bateson, Gregory. 1972a. "Comment on Form and Pattern in Anthropology." In Bateson, *Steps to an Ecology of Mind.* New York: Ballantine Books.

———. 1972b. "Style, Grace, and Information in Primitive Art." In Bateson, *Steps to an Ecology of Mind.* New York: Ballantine Books.

———. 1972c. "The Logical Categories of Learning and Communication." In Bateson, *Steps to an Ecology of Mind.* New York: Ballantine Books.

Benjamin, Walter. 1969. "The Work of Art in the Age of Mechanical Reproduction." In Benjamin, *Illuminations,* ed. Hannah Arendt. New York: Schocken Books.

Berger, P. L., and T. Luckmann. 1966. *The Social Construction of Reality.* New York: Doubleday.

Bjorken, J. D. 1976. "The 1976 Nobel Prize in Physics." *Science* 194 (Nov. 19).

Bohannan, Paul. 1967. "Concepts of Time Among the Tiv of Nigeria." In J. Middleton, ed., *Myth and Cosmos.* New York: Natural History Press.

Crane, Diana. 1980. "An Exploratory Study of Kuhnian Paradigms in Theoretical High Energy Physics." *Social Studies of Science* 10.

Eco, Umberto. 1979. *The Role of the Reader: Explorations in the Semiotics of Texts.* Bloomington: Indiana University Press.

Einhorn, Martin, and Chris Quigg. 1975. "On the New Narrow Resonances." *SLAC Beam Line,* March 20.

Ellen, Roy P., and David Reason, eds. 1979. *Classification in Their Social Context.* New York: Academic Press.

Evans-Pritchard, E. E. 1940. *The Nuer: A Description of the Modes of Livelihood and Political Institutions of a Nilotic People.* London: Oxford University Press.

Feynman, Richard. 1967. *The Character of Physical Law.* Cambridge, U.K.: Cambridge University Press.

Foucault, Michel. 1970. *The Order of Things.* New York: Pantheon Books.

Frake, Charles O. 1976. "How to Ask for a Drink in Subanum." In Pier Paolo Giglioli, ed., *Language and Social Context.* Harmondsworth, U.K.: Penguin Books.

French, A. P. 1968. *Special Relativity: The M.I.T. Introductory Physics Series.* New York: W. W. Norton.

Geertz, Clifford. 1973a. *The Interpretation of Cultures.* New York: Basic Books.

———. 1973b. "Thick Description: Toward an Interpretive Theory of Culture." In Geertz, *The Interpretation of Cultures.* New York: Basic Books.

———. 1983. "Art as a Cultural System." In Geertz, *Local Knowledge: Further Essays in Interpretive Anthropology.* New York: Basic books.

Giedion, Siegfried. 1969. *Mechanization Takes Command: A Contribution to Anonymous History.* New York: W. W. Norton.

Goody, Jack. 1968. "Time: Social Organization." In *International Encyclopedia of the Social Sciences.* New York: Crowell, Collier, and Macmillan.

———. 1977. *The Domestication of the Savage Mind.* New York: Cambridge University Press.

Haber, Heinz. 1956. *The Walt Disney Story of Our Friend the Atom.* New York: Dell.

Hacking, Ian. 1983. *Representing and Intervening: Introductory Topics in the Philosophy of Natural Science.* Cambridge, U.K.: Cambridge University Press.

History and Concept of Time. 1966. Special issue of *History and Theory: Studies in the Philosophy of History,* vol. 7 Middletown, Conn.: Wesleyan University Press.

Iser, Wolfgang. 1978. *The Act of Reading: A Theory of Aesthetic Response.* Baltimore: Johns Hopkins University Press.

Keller, Evelyn F. 1980. "Baconian Science: Hermaphroditic Birth." *Philosophical Forum* 11.

Kirk, Bill. 1974. "An Introduction to Colliding Beam Storage Rings." *SLAC Beam Line,* Aug.

———. 1975a. "The Positron Source Job." *SLAC Beam Line,* Feb.

———. 1975b. "End Station A Spectrometers." *SLAC Beam Line,* March 20.

——— 1977. "Preprints, Publications, and All That." *SLAC Beam Line,* July.

Kociol, Steve. 1970. "Does Time Run Backwards?" *SLAC News,* July 31.

———. 1971. " 'Vector Dominance' Works—or Does It?" *SLAC News* 2, no. 2 (April 19).

Lang, B. 1975. "Space, Time, and Philosophical Style." *Critical Inquiry* 2.

Latour, Bruno, and Steve Woolgar. 1979. *Laboratory Life: The Social Construction of Scientific Facts.* Beverly Hills, Calif.: Sage Publications.

"LBL, SLAC Designing Colliding Beam Accelerator." 1972. *Beam Line* 3, no. 3 (June 20).

Leach, E. R. 1966. "Two Essays Concerning the Symbolic Representation of Time." In Leach, *Rethinking Anthropology.* London School of Economics Monographs on Social Anthropology, no. 22. London: London School of Economics.

Lechtman, Heather. 1977. "Style in Technology—Some Early Thoughts." In Lechtman and Robert S. Merrill, eds., *Material Culture: Styles, Organization, and Dynamics of Technology.* 1975 Proceedings of the American Ethnological Society. St. Paul, Minn.: West.

———. 1984. "Andean Value Systems and the Development of Prehistoric Metalurgy." *Technology and Culture* 25, no. 1 (Jan.).

Leiss, William. 1974. *The Domestication of Nature.* Boston: Beacon Press.

Levi-Strauss, Claude. 1966. *The Savage Mind.* Chicago: University of Chicago Press.

Marcuse, Herbert. 1956. *Eros and Civilization.* Boston: Beacon Press.

Meyerhoff, H. 1968. *Time in Literature.* Berkeley: University of California Press.

Mulkay, Michael. 1979. "Knowledge and Utility: Implications for the Sociology of Knowledge." *Social Studies of Science* 9.

Needham, Rodney. 1979. *Symbolic Classification.* Santa Monica, Calif.: Good Year.

Olson, Richard. 1975. *Scottish Philosophy and British Physics, 1750–1880.* Princeton, N.J.: Princeton University Press.

Oxley, Charles. 1971. "MPC Improves Spectrometer Performance." *SLAC News* 2, no. 3 (June 2).

———. 1972. "LASS Construction Underway." *SLAC News,* April.

Pagels, Heinz. 1983. Plenary address presented at Conference on Science, Technology and Literature, Long Island University, Feb., New York.

Pocock, D. F. 1967. "The Anthropology of Time Reckoning." In J. Middleton, ed., *Myth and Cosmos.* New York: Natural History Press.

Richter, Burton. 1976. "Burton Richter: A Scientific Autobiography." *SLAC Beam Line,* Nov.

Ricouer, Paul. 1971. "The Model of the Text: Meaningful Action Considered as Text." *Social Research* 38, no. 3.

Schutz, A. 1975. *On Phenomenology and Social Relations.* Chicago: University of Chicago Press.

Sklar, Lawrence. 1977. *Space, Time, and Spacetime.* Berkeley: University of California Press.

"SLAC and MIT Collaboration Studies Proton Structure." 1970. *SLAC News,* Feb. 26.

Suleiman, Susan R., and Inge Grosman, eds. 1980. *The Reader in the Text: Essays on Audience and Interpretation.* Princeton, N.J.: Princeton University Press.

Thompson, E. P. 1967. "Time, Work-Discipline, and Industrial Capitalism." *Past and Present* 38.

Thornton, Robert J. 1980. *Space, Time and Culture Among the Iraqu of Tanzania.* New York: Academic Press.

Tompkins, Jane P., ed. 1980. *Reader-Response Criticism: From Formalism to Post-Structuralism.* Baltimore: Johns Hopkins University Press.

Tuan, Yi-Fu. 1970. *Man and Nature.* Association of American Geographers Resource Papers, no. 10. Washington, D.C.: Commission College Geography.

Vincenti, Walter. 1982. "Technical Knowledge Without Science: The Innovation of Flush Riveting in American Airplanes, c. 1930–c. 1950." *Technology and Culture* 25, no. 3.

Winsor, Frederick. 1958. *The Space Child's Mother Goose.* New York: Simon and Schuster.

Zenzen, Michael, and Sal Restivo. 1981. "The Mysterious Morphology of Immiscible Liquids: A Study of Scientific Practice." *Social Science Information* 21, no. 3 (1982).

CHAPTER 3

Engineering Design Process

Louis L. Bucciarelli

TRADITIONAL MODELS of engineering design, such as those found in the engineering textbook or professional journal, are themselves designs: rationally constructed plans developed by full members of the profession meant to describe how the design process works. Figure 3-1 presents one example of the generic "design process." Typical of most contemporary engineering texts, it lays out the process in block diagram form showing it as a progression through discrete stages, fifteen in this case, with allowance made for back-stepping or "iterating" around each phase; the arrowheads point up as well as down. The notion of different stages is common to most authors' renditions, some of which get very elaborate. For instance, the author of Figure 3-2 uses fully seven times more "stages" to describe the complex sequence of tasks that are performed in taking a design from inspiration to production.

As abstractions from experience, these images express an ideal upon which to pattern the design effort; they are a technical prescription for the temporal production of technique, meant to encompass all engineering design, all product development. Their instruction is control—control achieved by breaking the design task into discrete elements, sharply bounding each and every part, and then connecting them all sequentially with indelible, straight lines. As descriptions of process, they can be considered plausible accounts, expressing how many believe the process ought to work. They may also be useful pedagogically.

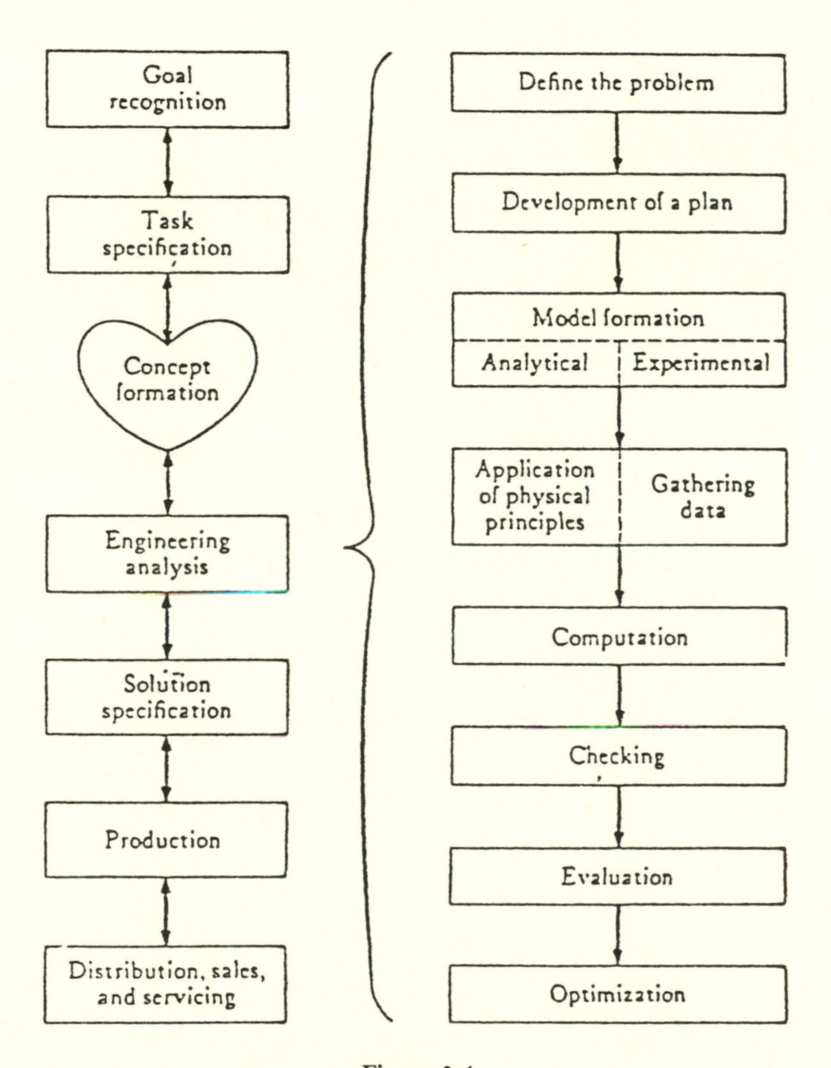

Figure 3-1
Block Diagram of the Design Process
Reproduced with permission from Dixon (1966), p. 11

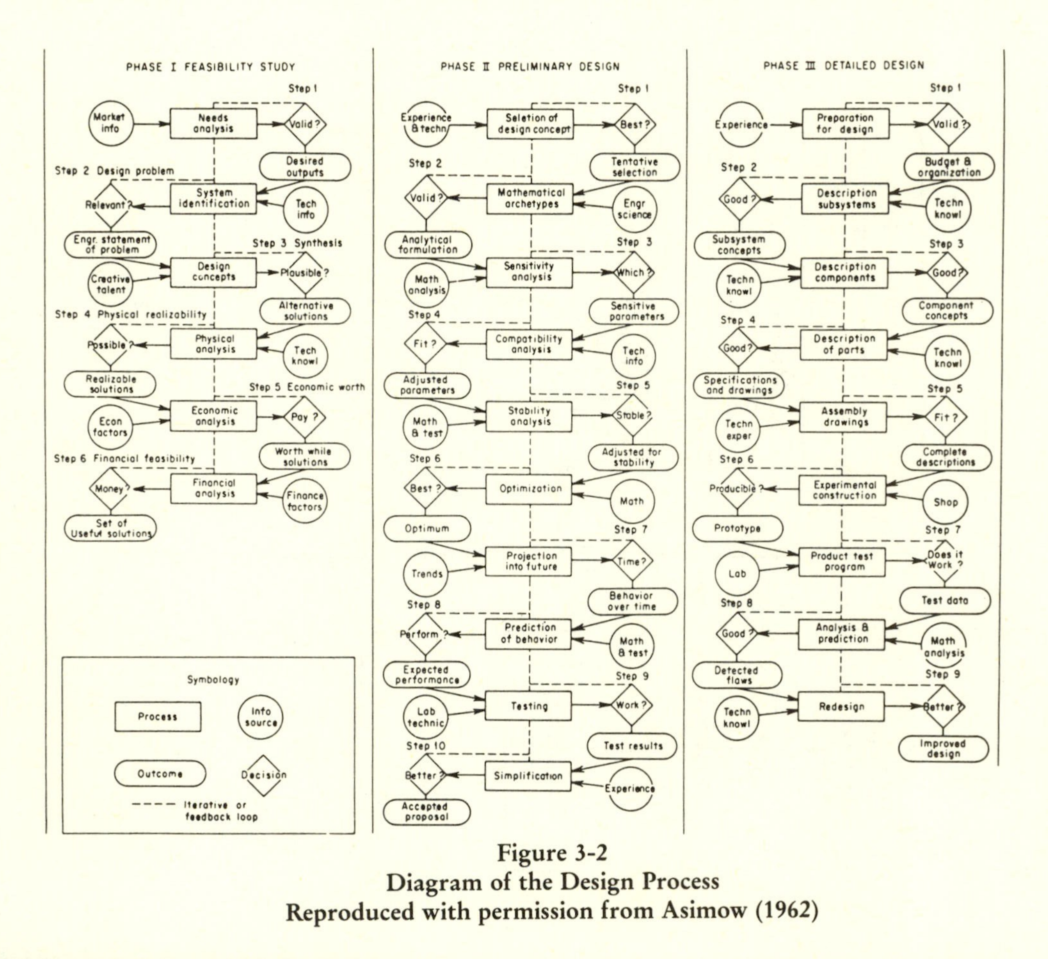

Figure 3-2
Diagram of the Design Process
Reproduced with permission from Asimow (1962)

Yet as models of practical design activity they are deficient. If we allow the figure to direct our thinking about those engaged in all the tasks contained in the boxes, we might conclude that design practice is an extra-orderly, rational process—one in which creative thought automatically leads to the fashioning of conceptual designs that, after detailed analyses, are given real form, tested, and finally put into production for profit as well as the benefit of all mankind.

By contrast, my account in this paper is based on a close-up study of design at an engineering firm—a firm engaged in making photovoltaic modules for the production of electrical power directly from the sun. I aim to get inside the boxes, not to draw other boxes *ad seriatum* as in Figure 3-1, but to construct a new description and analysis that rings truer, explains better, and might even prove more useful. My goal, as an engineer attempting an ethnography of his own profession, is to understand design as a social process, the engineering firm as subculture, and from this perspective to explore how technical constraints embody values and traditions, how these influence design decisions, and, to complete the circle, how technique itself—the world of objects, hardware, scientific law, instruments of design—structures values and norms.

My thesis is that to understand the design process it must be studied as a *social* process. The essence of this social process, however, is never really captured by the "native" models that govern thought and action within the engineering community—the ultra-orderly, rational diagrams of process shown in Figures 3-1 and 3-2. When these models are compared to the complex interactive texture of work in engineering design groups, they are seen to be clearly deficient.

Several questions arise: If these models are lacking, how do professionals deal with that deficiency? Now, it may be that Figures 3-1 and 3-2 are simply harmless abstractions of academics, which, while expressing an ideal, are of little use in the actual day-to-day world of design. Granting this, I claim they still reflect the way participants think design ought to flow. When it doesn't, when the sides of the boxes collapse, or when the entire image begins to distort and lose its shape, then something else must be going on. We must explain how participants in this imagined rational and orderly process accommodate that which cannot be shown using the machinery of block diagrams. Responding to this question, and others like it, will lead to a better understanding of design process.

A second question is more difficult: How will a "better" understanding of design, taken as a social process, be "understood" by members of the profession if it is not encompassed by, or indeed challenges the norms and values of, the culture?[1] Here the premise and hope is that norms and values can change.

A third concern, not unrelated to the last, is how to frame an analysis based on direct observation. I have elsewhere explored (Bucciarelli, 1984, pp. 185–190) the use of narrative, contrasted that with a topical structuring, and concluded that both can be employed to advantage. Narrative can claim the power to better convey a "feel" and sense of the world of engineering design while topical structuring suggests itself for dissection and analysis of that world. Here I go a step further and choose this binary format itself as a way to frame my study.

I posit two worlds within which design practitioners move: an "object-world" (topical) and a "process-world" (narrative). The object-world is the world of hardware, of performance specifications, of scientific theory and law, of quantitative estimates, of standard hex nut sizes, of budgets and milestone charts—all the instruments and machinery in mind and in hand that are brought to bear in design. It is the world of the participant engaged with the materials of design. The process-world is the world of dialogue and negotiation, of social exchange, laughter, gossip, banter—all that which is ever-present in design, but whose significance is generally discounted. Object-world thinking gives us the models of design process shown in Figures 3-1 and 3-2. Attending to the process-world gives us a way to understand better the essence of design as a social process, and hence the deficiencies of that other too narrow and too instrumental perspective.

I focus on "time" as a dimension within both worlds, as a thread that weaves through both but appears in different forms, has different meanings, depending upon our perspective. In the object-world, the significance of time is reflected in a variety of instruments used in the planning of design tasks—the milestone chart, the CPM chart. There it is measured; there it flows continuously and uniformly. In the process-world, its character is revealed in conversation, in often informal and apparently trivial discussion. There it is not quantified; there it starts and stops, rushes by at times, drags at other places in the process.

This dual nature of time, time objectively measured, time subjectively experienced, is a common enough notion. What makes it worth pursuing here is that design participants impose their objective constructions upon the world of process to a degree perhaps unrivaled by any other profession. The object-world cosmology of engineers and managers dominates in design and the traditional tension between "time objective" versus "time in-process" is tremendously heightened. How this tension is accommodated is itself an important factor in creative design.

OBJECT

One world of design appears as an objective reality—the world of sharp lines on a drawing, of written codes and specifications, of "hard" laws of science and empirical fact. Within this object-world, time is both a resource and a reference. As duration, it is a resource to be meted out, distributed, conserved, spent, or wasted away. As such, it is finite; at any time there is only so much of it left before the next deadline. It can be measured with precision, whether the metric be years, as in the mean time between failures of a photovoltaic module, or in microseconds, as in the duration of response of a voltage step within a microprocessor controller. Small or large, it is all the same stuff, neutral, weightless, colorless, but valuable.

Time as resource is exemplified in another object-world diagram, one actually used in design: The milestone chart of Figure 3-3 is an example. It can be viewed as a snapshot of a month in the life of a participant in design, a picture of how his or her time is to be "spent" over the next one or several months. Tasks are listed in the column at the far left; time, measured in increments of days, runs horizontally to the right. Estimates of the number of days required for each task are also shown. The engineer to whom this chart belongs has finished work on a Pumping Application Note. He is currently polishing up the final draft of the user's manual for the Jeddah Clock Tower Monument, an effort required in the final stages of design. Much work remains to be done on the design of the state of charge controller for the desalination system for Qatar. Finding replacement parts for the desalination plant at Jeddah is an ongoing process. Training the technician from Jeddah will consume most of his time over a two-week period. The test set-up for evaluating various manufacturers' voltage regulators has been put in place and testing has begun. Preliminary designs for a pumping system for Haiti and another for Sierra Leone have recently been completed. One for Gabon remains in the works.

The chart suggests that there exist clear and distinct beginnings and ends to design tasks. What can be surer than a "deadline"? Duration comes in finite chunks. Just as the block diagrams of design process imply that each box will take up a certain amount of time, so too the horizontal distance on the milestone chart represents, but in this case explicitly, a set amount of time. Note, however, that in both examples the calendar dates of the start or end of any period, the "at this point in time" times, are left free and open. In the case of the milestone chart, the user fills them in as need be. The block diagrams of process, on the other hand, are meant to stand free of time; as other models of process in engineering and science, they stand "timeless," true for all time.

Still other charts, while not accounting for time, give the same vision of permanence, certainty, and order. These hold time still and display the way "design space," in the sense of "state space," is structured.

Organization charts, for example, define the state of the enterprise from a managerial perspective. They come in different levels of detail at different levels of the firm, describe hierarchy and the variety of functions individuals or groups perform. The president appears at the apex of the most global image of the photovoltaic firm. The heads of each of the five groups beneath him, Research, Production, Marketing and Systems Engineering, Comptroller and Administration, and Employee Relations, branch off below. Within each of these five groups there are further subdivisions. More tree-like diagrams can be and are drawn that show finer details of the organizational state of the firm just as Dixon has drawn up, within one of the boxes in the figure depicting the design process (Figure 3-1), another string of boxes.

Plan views of the layout of office and laboratory space are similar. They too influence the organization of design and show the state of the firm segmented: each group's turf sharply defined and bounded, then further broken down into compartments and pockets where individuals have control over the environment and over the conduct of discourse. Discontinuity, as well as sheer size, is valued, as

Figure ______ Milestone Chart For ______ Project No ______

		AUG.	SEPTEMBER																					OCT.	
Task	D	31	1	2	3	6	7	8	9	10	13	14	15	16	17	20	21	22	23	24	27	28	29	30	22	29
Voltage Regulator																										
Design Test Set-up	2																									
Build Test Set-up	3																									
Testing	10/21																									
Manufacturers Data Input	4																									
Haiti Pump System																										
Parts, Pricing & Ship																										
Date	1/2																									
Sierra Leone (Africa)																										
Lighting, Refrigeration																										
Fan, 2 Pump Systems																										
Parts, Pricing & Ship																										
Date	'2																									
Gabon Proposal																										
Review/Comments	1/2																									

Prepared by ______
Date 9-3-82
Rev A Date 9-22-82
Rev B Date ______

Figure 3-3
Milestone Chart

Figure ________ Milestone Chart For ________ Project No ________

		Aug.	SEPTEMBER																						OCT.	
Task	D	31	1	2	3	6	7	8	9	10	13	14	15	16	17	20	21	22	23	24	27	28	29	30	4	19
Pumping App. Note																										
Final Review																										
Update Draft	1/2																									
Prepare for Printers	1/2																									
Jeddah Clock																										
Prepare T.S. Table	1/2																									
Expedite Manual & Parts	2																									
Pump Model																										
Test & Approval	1/2																									
Ship (Fayhey)																										
Qatar																										
Test CTA	1/2																									
Test SOCC	8																									
Jeddah Desal																										
Find Replacement Parts																										
John G																										
Training																										

Prepared by ________
Date 9-3-82
Rev A Date 9-22-82
Rev B Date ________

Figure 3-3 (cont.)
Milestone Chart

indicated by the variety of environments I encountered—a variety that often expresses relative status.

The object of design itself—for example, a new photovoltaic module—shapes the design space. The synthesis of all the different parts and functions of subsystems of a module suggests patterning the organization of the design task on the artifact itself. The different functional subsystems provide a useful way to structure the relationships among participants in design. Sometimes the play of this kind of organization against the organization of the firm is itself given formal expression in another chart, a matrix showing participant-task associations.

The aim is to break up the design task so that each participant works within his or her specialty, within a small group, a sub-sub-culture within the firm, and that group's work can be pursued independently of all others. Integration of the different groups' efforts and designs is scheduled, of course, in a planned sequence of steps. The interactions between groups are expressed and controlled through a statement of "interface specifications." Like the charts of Figures 3-1 and 3-2, the formal ordering of design tasks, breaking them down into smaller tasks, in theory like any taxonomy, can be carried on endlessly.

The organization chart, plant layout, project breakdown schemes, and the design text's figures of process all serve as reference maps for design activity. A place for everything and everyone, and everything in its place. Again the intent is control, the setting of a foundation for the management of process, valid for all time, or at least for the duration of the project.

Time likewise serves as a reference for gauging change within object-worlds, serving as a *background dimension* against which one accounts for movement and development. The block-diagram of the design process suggests a uniform flow of time from top to bottom. But if "feedback" or "iteration" demands that one go back a step, the clock keeps ticking, time continues to roll by.

Often time flowing is described explicitly in technical terms. A senior executive describes the problems of coordinating different, more or less independent, design activities under his jurisdiction—for example, the development of a piece of the technology of photovoltaic module production, the design of a new photovoltaic module, and the introduction of an innovation in cell processing—as a problem of "getting things to move in sync. . . . We have problems if things get out of phase." This is the language of control theory. The image is of developments evolving independently as time goes by at a steady pace. The activities themselves move slower at times, speed up at other times. They must all come together at some common time if they are to be productively integrated.

Another effort at time management, like the milestone chart, that is applied in attempts to keep a hold on diverse but interrelated activities is the Critical Path Method chart, a graphic display of the sequences of steps that must be followed, and how they must come together, to ensure that a design deadline can be met. Like the text's image of the design process, the CPM chart focuses on the artifact, the thing being designed. These charts are about "things" and define key events, such as when supplies appear, when they are integrated into the design and sent on, when a subsystem must be delivered, when the product must be shipped out the door. Looking at the network of interconnected circles on a CPM chart you see the artifact coming together—like the image conveyed by running the film of a shattering accident backwards.[2]

Again there is a finite time span from start to closure. But the computer algorithm that takes in the data and calculates the critical path, the separate branches of flow, and estimates the allowable duration of each step, is insensitive to calendar date. A final delivery date can be chosen freely, but once it is chosen the machinery of the algorithm defines the starting date.

The milestone chart and the CPM chart show duration

measured and manipulated in finite and independent chunks. So too the organization chart, the layout of work, of lab and office space, and the compartmentalizing of the design task into subsystems and specialties show the state of design broken up into finite and independent "spaces." These elements in this reductive enterprise appear simple, basic, and atomistic—a direct inheritance of seventeenth-century Cartesian thinking. From within the object world, time and space are flat, neutral, in some ways uninteresting, but extremely important design dimensions.

PROCESS

We turn now to a consideration of process and another perspective upon these elements to develop a fuller interpretation. The process of design is the marketplace where design is negotiated, where ideas are exchanged and decisions made. We leave the world of the designer and his or her primarily solo manipulation of objects, ideas, images and plans and take a broader view. Now time has a different nature; in the world of process no measure is made, no metric applied. The length of time to frustration when the computer "goes down" would read, objectively, in minutes; the experience in process has a totally different quality. A measurement to a time scale is meaningless (except as a statistic to an object-world analyst).

Consider the beginnings and ends of projects: When does the design process start; when does it end? The milestone chart and the CPM chart suggest that tasks are all of finite duration and are bounded by well-defined starting dates and deadlines. Within the world of process it is not as clear. One could safely claim that the the design of the firm's original photovoltaic module began within the last two decades, for twenty years ago the firm did not exist. But even then, that particular product had its antecedents. With more precision I note that one year ago the firm marketed just the one product. Today (1986) it markets a wide range of modules. So now, while being phased out of production, the original product serves as a precedent and pattern for new designs.

To mark the origin of a design of a particular product is impossible: a request for quotation, a corporate directive, an idea in the lab, a "natural evolution from last year's model" promoted by Marketing, all these might be posed as the first step in the process. They are far from simultaneous events. Fixing when design ends is likewise difficult. The design is "frozen, "a prototype has met most expectations, product specifications are released to the market in full gloss, significant sales have accumulated, all these can be argued to mark the completion of design.

The origin and end of design cannot be labeled with a specific date or defined by a specific action. Beginnings and ends are diffuse, though not necessarily in the eyes of any one participant in design. But because the design process engages a variety of people with different perspectives and interests, each will have a different view of beginning and of end. Consensus is not possible; agreed-upon beginnings and ends to a story, precise markings in calendar time, do not exist (except in the historian of technology's reconstruction).

The boundaries defined by the organization chart, by the layout of the firm's actual physical space, and through the definition of the artifact's subsystems, though appearing sharp and fast when drawn on paper, are just as diffuse and fluid. The formal organizational boundaries do indeed define the home turf of group members but the walls are quite porous. There is exchange across the lines prompted by work: Shipping connects to Marketing when deadlines given a customer are going to slip by; Module Production frequently negotiates with Systems Engineering on some details left ill defined at the formal design review; the computer goes down, a promised phone line is a month overdue, an employee benefit is readjusted, and the direct exchange across empty space between the boxes on the charts intensifies. Other traffic is social. Note that the charts suggest these excursions are out of line.

From within the group's space, the horizon presents a more varied profile than the near field described above. Beyond the other groups and individuals within the firm, there is corporate headquarters in New York City, there are the firm's distributors, there are subcontractors and suppliers and travel agents, and at precisely 11:20 A.M the bell rings twice sharply, signaling the arrival of the canteen man. Corporate executive, Alaskan distributor, solid-state physicist, plumber, Arab sheik, lawyer—if an agent of the U.S. Department of the Interior walked in with a request for proposal for a photovoltaic system to electrify a Zuni village in the Southwest it would not be out of context. The space of participants in design is varied, neither homogenous nor the same in every direction.

All this rich detail, all these channels of communication and exchange both within the firm and reaching out into the world, fail to be reflected in the formal organization chart. They connect to regions not under the control of the firm and, hence, are the occasion of uncertainty in process. But knowing how to use them is as essential to design as knowing the proper routing to place on an ECO, an engineering change order.

The permanence of the organization chart is similarly misleading. It is not as stable in time as its form suggests. Changes in the names of the occupants of the boxes, changes in the titles of the boxes, and even changes in the layout of the functional units are not infrequent over the course of a year or two. Still, at any instant, a participant's place is well defined, at least formally. Occasionally there will appear a dotted connection representing an infrequent alliance, a gesture of "will keep you informed," or uncertainty with what to call a hard-to-place but useful person.

The milestone chart offers the same illusion of definiteness. While it suggests the continuous chunking away of a finite number of days to come, from the perspective of the individual whose milestone chart it is, the exercise of its construction has an element of fantasy about it, asking for too high a degree of precision in pacing future, uncertain events.

To account for one's future in the terms of the chart engenders an uneasiness, a sense that its format is too confining and disallows any adequate explanation of what it will take to get the job done. Time is not all that is required. Like the text's account of the design process, this object-world instrument is deficient.

The same observations apply to the formal attempts to compartmentalize the design effort into smaller, more manageable, better-defined subtasks guided by the features of the artifact itself. The segmenting of design appears to make rational what can be a very uncertain operation, uncertain because the interfaces between subsystems are not given *a priori* but must be constructed.

For example, in the design of one new, large module, the Panel Fabrication group has responsibility for the wiring of the modules out to two stud connectors. Systems Engineering has responsibility for the junction box that fits over the studs and includes posts for the connection of diodes and external wiring. An interface is defined by the geometry of the studs, and the need of fixing the junction box to the module. But these spatial, or geometric, interfaces are not the only ones. The design of the junction box must take into account the heat dissipated by the diodes within the box. This, in turn, depends on the electrical characteristics of the network of cells, which is the responsibility of Panel Fabrication.

Interfaces then cannot be set uniquely to satisfy all participants in design. You cannot draw a single envelope around a subsystem that best isolates the component from the perspectives of different technical disciplines. Deciding who should have responsibility for what tasks, deciding where to draw the boundaries, is not unambiguous. The junction box could have been the responsibility of Panel Fabrication. This interlacing of specialties is one reason why, just as in the organization chart and the plant layout, the boundaries between design tasks, represented as precise, are, in process and practice, diffuse.

Despite the appearance of continuity and order conveyed by the charts, the generally articulated perception of time is that there is never enough of it. All the machinery of scheduling, forecasting, and systems analysis can never fully define, hence control, the future. Uncertainty, no matter how comprehensive or detailed the charts become, is always in the air. Indeed, there is a limit on how much designing of design is worthwhile. The planning effort presents the same problems it is intended to solve.

But there is a sure flow and direction to the process. The design of the firm's newest product, the large photovoltaic module mentioned above, initially had the attention of but a few key persons—in Marketing and Systems Engineering and Panel Fabrication. As time passed, the design engendered further commitments, and demanded the accumulation and expenditure of the resources of others. The larger module required new tooling and fixturing, so production people had to be brought into the act. Others had to be trained in the peculiarities of the assembly process. As specifications were drawn out, suppliers had to be found who were willing and able to supply materials and components on time. Analyses were conducted to define new structure, to determine reliable cell networks, and to estimate performance. Accounting and Purchasing and other service groups had to learn a new vocabulary.

Eventually, all groups within the firm were touched by the design task and became part of it. I am reminded of the TV commercial that starts with the image of a ten-year-old, mud-stained lad opening a bottle of pop. He hums a catchy refrain, and then down the street is joined by his mom, the neighbors, the milkman, the policeman, his dad, and finally a barking dog and the high school band. So too the flow of design shows an accretion of detail in stages, even widening in its demands and influence. If I were to attempt to jazz up Figure 3-1, I would multiply the boxes, increase their size, and add some gothic ornament as one goes down the path.

While different participants in different positions in the parade perceive the evolving design differently, all value continuity in their association with the task. But keeping hold of the pace and rhythm of a project is not automatically ensured by a consistently strong attendance record. Every participant, every group, works on a variety of projects at any moment. Design task A will call for effort one day; then B will come to the fore and require attention, pushing yesterday's project out of view. When return to A is possible, days, a week, a month, may have gone by. Reengaging the old project can be frustrating, since over the course of time, the framework of design flexes and shifts, the context of the design task is altered, and, in particular, constraints and specifications may have changed.

These discontinuities can be a minor nuisance requiring some, but not overbearing, effort to accommodate, or they can reflect a more substantial change in the path of design and require retrenchment, re-evaluation, and often the discarding of ideas and plans that had attracted considerable investment.

While the text's account of the design process shows that reassessment is necessary through a "feedback" process, it says nothing about the multiplicity of demands on, or the potential for disruption of, the lives of those attending to the tasks defined at each stage in the figure. Delays, which in retrospect are seen as avoidable but which, in process, and not sorted out so easily, occur, due sometimes to the most unimaginable reasons—a supplier goes bankrupt, a dock strike takes weeks to resolve, a colleague breaks her arm.

As the different design tasks peak and subside, some surface with a special urgency, bursting upon the scene as "a fire to be put out." It is these experiences that prompt the hypothesis, "If there is a possibility of several things going wrong, the one that will cause the most damage will be the one to go wrong." Murphy's Laws are prominently displayed on the walls of every design world. Like graffiti, they encapsulate truth, albeit truth reduced to the absurd. Many of them deal with time and uncertainty. However flip they appear, they can be understood as attempts by design cultures to explain that which cannot be accounted for in object-world terms—the world of design process as social process.

CONTEXTS FOR DECISIONS

There is a flow to design: hesitating, appearing to sometimes falter, at other times to accelerate out of control, but still directed forward by participants toward the final fixing of ideas, constraints, and values in the material form of an artifact. Occasionally design flow is suspended and organizational boundaries ignored; participants meet on higher ground to clarify and reconcile different design ideas, to generate and evaluate fresh alternatives. In short, decisions are made and the flow and structure of design are reaffirmed or altered. Decisions made in this setting I call "hard"—decisions made at formal meetings, gatherings that may be small, with only two present, or large, with the entire group present, but meetings scheduled with an agenda, though it may not be adhered to.

Decisions come in forms other than hard and formal. A happenstance gathering in the hallway, a background conversation at a group meeting, an intense dialogue at a farewell party, and the like can be settings for decisionmaking. In the course of casual conversation, limits to design thought and practice are established. A verdict may not take written form—indeed it may not even be explicitly articulated, and agreement may only be signaled by a shared burst of laughter or general assent to a participant's choice of word of condemnation—but choice is made in this way. These "soft" decisions define the context of design, fix what alternatives and ideas will be entertained, which will be considered laughable or ignorable.

For example, at the meeting where the milestone chart of Figure 3-3 was discussed, the following exchange took place:

> *Systems Group Leader:* How goes the qualification of the regulators?
>
> *Engineer:* . . . Nothing much has changed.
>
> *Systems Group Leader:* Are your milestones still accurate?
>
> *Engineer:* Hopefully . . . but I've only put a half-day for test set-up. . . . Hopefully the whole thing will be finished towards the end of November."
>
> *Systems Group Leader:* November is a long way away. . . . It's got to be done. . . . Every system has a regulator.
>
> *Engineer:* I'll get it done by then.

In fact, every system does not have a regulator; a voltage regulator is *not* an absolute necessity; not all photovoltaic systems in service out in the field have one. The Systems Group Leader is saying, in effect: "Our designs must include one." The pressure applied to one of the engineers in his group to accelerate his qualification-test program establishes and reinforces that design constraint. The deadline date, set at the end of November, is not critical.

Hard decisions, on the other hand, decisions explicitly

acknowledged as such, are said to be made by resolving conflicting and competing alternatives, testing them against some generally accepted technical constraints. Constraints do indeed help define the artifact, but they are far from decisive. One might claim that there are always other alternatives that could meet these constraints, and significantly different alternatives at that—for example, alternatives that imply a significantly different social use. But it is impossible to win that game. Positing different consequent designs given a set of well-defined constraints will always remain, no matter how reasonable the alternative designs appear, mere speculation. I believe it more instructive to show how constraints themselves are constructed in the design process. This puts the cart back behind the horse where it belongs, for constraints are far from "hard" *a priori,* but rather are defined within the flow of design, evolving with the artifact.

For example: At the outset of a design effort, in splitting up the design task into subtasks and systems, when the terrain on all sides of all interfaces is largely unknown, it is impossible to construct a complete set of specifications valid for the life of the design. If a comprehensive set of specifications is established, they still remain open to the challenge in process. Creative design continually questions the rules of play. There is a tension between wanting to fix specifications at an interface for all time and the thrust of innovative design to overrun or at least challenge the rules.

Because it is impossible, and generally not desirable, to fix all the conditions at the interface and because they are open to challenge in the course of design, it is not uncommon to see conflicting designs emerge. In their resolution, fresh interpretations of interface conditions evolve, and more detailed specifications are adjoined to the project. In this way design proceeds.

Design is more a matter of shaping perceptions than satisfying given constraints. That is why soft decisions are as important as hard choices. Different participants have different images of the objects of design. There is a good bit of uncertainty and many loose ends associated with these. At the start there are sparse images of, say, a junction box. There are images of junction boxes past, of those of the competition, of ordinary household varieties. Some form images of materials and their thermal and structural properties, of how costly it might be to have someone tool up and fabricate an order. Others have images of appearances—how the "j-box" will look appended to the backside of the module, or whether it should be black or clear. Still others focus on how it will be assembled and how connectors will be attached. These images vary with the flow of design.

As design proceeds, perceptions change and emphases shift. Yesterday's important considerations may no longer be today's critical issues. This too depends on whom you ask, his or her responsibility and concerns. Yet consensus must be built and maintained, though not on every detail of every item of design. Rather, participants must agree on what are the significant questions and problems at any stage, on the limits of legitimate alternatives, and on the level of detailed analyses, exploration, and justification of options that must be engaged. The task of design then is not just to add flesh to the bones but to ensure that everyone is working with the same skeleton.

The shaping of consensus and perceptions can be informal and casual or, at times, can take the form of a negotiation. When milestones are the topic of a Systems Engineering Group meeting, a negotiation between group leader and chart maker takes place. (Others present contribute as well.) Questions are raised and estimates of time allotments defended. The means and resources required to complete all the tasks are clarified. The chart is then a shopping list of what is to be purchased with an individual's time; and constructing the list, negotiating appetites, is shaping common visions of the ingredients of a design.

The milestone chart and the other diagrams and instruments are abstractions whose intent is to facilitate agreement on the way participants as well as subsystems and elements must relate and flow through time for design to succeed. They are like maps—subway maps, contour maps, street maps—all covering the same terrain but focused differently. But although they appear as plans of how hours and resources are to be neatly spent in the future, they are rarely referred to once the meeting is over. Their construction, their social construction, is what is significant to the design process. After the meeting, the chart remains in the desk drawer or on the bulletin board for reference, but it is usually superfluous for day-to-day travels. Its importance lies in its creation.

Here is its value in use. Indeed its making is a significant phase of design activity itself and a context for decision making. In the process of its construction, choices are being made. The planning of the design task through a categorization of the artifact-to-be, this *a priori,* general dismembering, the layout of office space, the specification of organizational responsibilities, all of these channel thought in some ways and not others, allow for certain design configurations and not others. In these object-world acts the artifact is prefigured, decisions have been made.

DESIGN DECISIONS AMIDST AMBIGUITY

Consensus does not mean full agreement on every detail of design or that everyone shares the same perceptions on the problem of the moment. In fact it appears that there should be a certain amount of ambiguity in the minds of individual participants doing decision making, especially if the decision is required among different interests.[3]

An account of a meeting called to set the design of the junction box for the large module illustrates the point. At the outset, what seemed like an unnecessarily lengthy amount of time was spent in meandering conversation ranging over what at first hearing sounded like unconnected aspects of the design, not of the junction box alone, but of the entire module.

> *Systems Engineer #1:* What cell size are we going to go with?
> *R&D Group Leader:* What have we promised our customer in the way of module performance characteristics?
> *Systems Engineer #1:* We're verbally committed to a 70-volt module but that's not set in concrete. . . .
> *System Group Leader:* When did we promise delivery?
> *Panel Fabrication Leader:* When will the glass superstrate be shipped to us?
> *Systems Engineer #2:* What's the overall module size. . . . What are its external dimensions?
> *R&D Group Leader:* Let's get back to business. . . . Where will the diodes be placed? We need an engineer's layout. . . .
> *Systems Engineer #1:* . . . Can we integrate them [diodes] into the module . . . ?
> *Systems Engineer #2:* . . . JPL sets a spec at one every 12 cells.
> *R&D Group Leader:* What module voltage are we talking about?
> *Systems Engineer #2:* . . . If we went with two junction boxes. . . .
> *Systems Engineer #3:* . . . A 36-volt module would be ideal for every 12 cells. . . .
> *Systems Engineer #2:* Where do we locate the junction box? . . . How do we tie it down? . . .
> *Panel Fabrication Leader:* . . . Are we going to have a frame? . . . If we build a 12-volt module we won't need diodes at all. . . .
> *Systems Engineer #1:* How many cells in series to get the 12 volts?

These questions are interrelated. Each embodies uncertainty. Each sparks and is sparked by dissimilar images in the minds of different participants. To those in Panel Fabrication, the panel (not the module) is a collection of fragile photovoltaic cells—not an arbitrary collection but a set of matched units. They know the cells' histories, how they were sorted, cultured, doped, and laced with conducting filaments. A panel brings to mind the careful treatment accorded these slender slips of crystalline silicon in laying them onto a panel of glass, connecting strings of them together in an electrical network, protecting them from the weather with a substrate of layers of foil and tedlar.

To those in Systems Engineering, on the other hand, the photovoltaic module (not the panel) is itself a unit, an elementary basic building block of a photovoltaic array of considerable power. They too see a stylish object, but to them it is an artifact that, when exposed to the sun, produces electrical energy directly. Their module sits on a rooftop on an island off the coast of Maine or in a rack with others in the desert of Saudi Arabia, and draws its meaning from the performance of a system of which it is only one part, albeit the most expensive part.

About the only commonly shared belief about the module before the junction box meeting was that one of its overall dimensions was to be 48 inches, to match with the standard construction practice, and that the output of the module would be fed into an inverter and transformed from direct to alternating current (DC to AC). All else was subject to negotiation, even the necessity of having a junction box at all. Answers to all the other questions were to be made, not found.

In time, discussion focused on the question, What should the module voltage be? Other questions might have served as well to channel negotiations, but the voltage was construed, constructed, as the problem. Ambiguous enough to enable the maneuvering of presumptions and latent images, it served as a cover term for all questions.

Just as different participants have different visions of

what a module is, so too their views vary about what a new module should be. Systems Engineering wanted a high-voltage module, Panel Fabrication a low-voltage module. A high voltage would allow the Systems people to match the requirements of the DC-AC inverter with fewer modules. Marketing advantages were also claimed, and economic studies showing this advantage were cited. Furthermore, a government laboratory with responsibility for the development of low-cost photovoltaic technology wanted to test a large module.

Panel Fabrication noted that a high-voltage module would require by-pass diodes and it was unclear how to network the cells and arrange the diodes to get the high voltage that Systems wanted. Furthermore, just what high voltage was desired? A 12-volt module could be achieved without any diodes in a simple, unique, straightforward way.

This uncertainty over what high voltage was required, and the ambiguity surrounding what was implied by any particular value chosen, provided the space for consensus—not just on the voltage question but on the full array of questions posed at the start of the meeting.

The Systems Engineering and Marketing Group manager recommended 48 volts. Two other members of the group worked out, on the board at the front of the room, how the cells could be connected in series and parallel to get this voltage. An appropriate, satisfyingly small, number of diodes seemed to spring from this sketch with little additional effort. Junction box placement followed just as easily. The whole pattern became clear, priorities were established, there was banter and laughter, then closure and scheduling of another meeting to firm up the details.

After the meeting, as I was questioning key participants, it emerged that different individuals had quite different interpretations of the significance of the choice of a 48-volt module, both in terms of the process by which they arrived at that number and in terms of its technical implications. Regarding the former, Panel Fabrication, who had wanted a 12-volt module, interpreted the choice as a compromise: "Systems orignally wanted a 96-volt module." Their main concern was that the diodes and the necessary network of all interconnections could be conveniently accommodated in the lay-up of the module. Eight diodes had been selected at the meeting, not excessive at all, and the interconnection scheme appeared efficient. (Subsequently, the number of diodes was increased to 12, but by that time the maximum module voltage was set; it was no longer negotiable.)

Systems Engineering claimed that what was chosen "was really a 60-volt module . . . at JPL standard testing conditions." The *nominal* voltage, 48 volts, was just that—a name, a label, serving as a "key symbol" (Ontner, 1973, pp. 1338–1346). Such key words or symbols derive their utility from lack of precision and the room they provide for different, even divergent meanings. The word "force" or the term "energy" in classical mechanics historically shows this ambiguous richness. Terms such as "module voltage" or "efficiency" offer within their own limited domain the same opportunity for creative design of photovoltaic technology.

Even though the "objective" engineering layout of the network of cells, diodes, and cell interconnections had been clearly sketched on the board, and though everyone "saw" the same object-world representation, different participants had different interpretations of the implication of that image. They differed in terms of module performance, ease of assembly and fabrication, costs, customer appeal, and the like.

The orchestration of the design process calls for a subtle flexibility. Not that one should attempt to foster ambiguity among the members of the group, but in the face of the kaleidoscope of constraints, knowledge of specialties, partisan, often conflicting interests, and, just as important, the uncertainty of it all, the laying on of more rational formulation and detailed block diagrams appear as weak instruments indeed. Yet making hard decisions is necessary at each stage. Participants recognize that design elements must be specified with some degree of firmness as the design evolves, for drawings must be made, fixing ideas and enabling costs estimates to be obtained. On the other hand one doesn't want to go out on a limb unnecessarily and prematurely "freeze" elements of design before one has explored as many alternative concepts, materials, and methods as possible. This dilemma becomes more acute with time, as concepts and subsystem designs become more intertwined, as deadlines and design reviews become more critical, and as resources committed to the design task increase.

The art of design consists in knowing what degree of specificity is required at each stage. All decisions have an element of irreversibility; all condition the design context and limit the range of options of others. Similarly, others' interests and specialties are contingencies. Through it all, everything appears in flux, interwoven and turbulent. Still, the shutter must click, images must be fixed, decisions made.

CONCLUSION

Time and "space" are two dimensions of the engineering design process. Within object-worlds they are like the coin of the realm—basic, atomistic, additive, and of uniform value. Within process-worlds their independent status dissolves and they show different traits.

A diagram, meant to summarize how object-world and process-world relate, is useful:

Object-World	*Process-World*
• Time of finite duration	• No beginnings, or ends
• Space with sharp boundaries	• Fuzzy, diffuse boundaries
• Time uniform	• Time colored by context
• Space permanent	• Space in flux

The claim here is that ambiguity and uncertainty are essential to design. Attempts to banish them completely are not only doomed to fail but would choke the flow of the design process, bringing it to a lethargic, dispiriting halt. Ambiguity is essential to allow design participants the freedom to maneuver independently within their object-worlds save for ties to others to ensure some degree of consensus. Ambiguity is found in different participants' perspectives in design, in their singular interpretations and understandings of terms. This is necessary for design to happen at all. "How much" ambiguity is tolerable, or is optimum, is an object-world question that, if answered, would drive the beast from the forest. No measure is possible.

Uncertainty too is always present. It is what gives life to the process, what makes it the challenge that it is. If design proceeds without it, something is wrong, the process is not designing but copying. Now it may be that we can construct a computer heuristic that can, given a field of constraints, search and find a "satisficing" design. This dull task would not be designing. The buck would pass back to those responsible for the design of the computer heuristic, to negotiation and battle on a new field of constraints. There uncertainty and ambiguity will abound as before.

The main value of the object-world instruments used to assist in design lies in their construction. The social process that produces them is a way to deal with ambiguity and uncertainty. No instrument in itself is sacred, or is better than any other simply on the basis of its apparent machinery. The context and process are all-important. That context is social.

NOTES

Acknowledgment: This paper is based upon work supported by NSF Grant No. ISP-8114659 in the NSF Ethics and Values in Science and Technology Program. The author thanks Sharon Traweek, Frank Dubinskas, and Don Schon for their guidance and constructive criticism.

1. Even if the new account is appreciated or applauded, even published in a minor journal, the question still remains if there is no "understanding" in the sense that it fails to affect the thought and actions of practitioners in design, doing design.

2. Film is an appropriate metaphor: Object-world time can flow by uniformly and continuously as in a theater. But a film editor in the studio can stop the flow at any point, look frame by frame, start and stop, review again as needed. The editor controls the flow, manages time. So too the design practitioner attempts to control the elements and stages, the flow of the design process. Process-world time is a dramatically different experience but still the metaphor applies, except now the design practitioner is in the film, seemingly unable to alter events as they surge past.

3. "But the impact of uncertainty is not always a cost; it can occasionally actually improve an outcome. Administrative action is not simply a matter of deciding what to do. Decisions often result from lengthy processes that involve all sorts of interpersonal wrangles and Machiavellian tactics. Uncertainty can help to bring these processes to the point where some measure of agreement is achieved. Given the ambiguities and cross-pulls of political life, given honest differences in values and factual judgments, I wonder how often people would agree on a course of action if everyone knew precisely what they were agreeing on. The uncertainty inherent in all aspects of decision can provide the leeway or a rearrangement of fact and emphasis which makes coalition possible and a strategy of achieving consensus effective. The uncertain world must fight fire with fire. In this sense, the obfuscation of uncertainty can be an advantage rather than a cost" (Mack, 1971, p. 6).

REFERENCES

Asimow, M. 1962. *Introduction to Design.* Englewood Cliffs, N.J.: Prentice-Hall.

Bucciarelli, L. L. 1984. "Reflective Practice in Engineering Design." *Design Studies* 5, no. 3 (July).

Dixon, J. 1966. *Design Engineering.* New York: McGraw-Hill.

Mack, R. O. 1971. *Planning on Uncertainty.* New York: Wiley-Interscience.

Ortner, Sherry B. 1973. "On Key Symbols." *American Anthropologist* 75.

CHAPTER 4

On Technology, Time, and Social Order: Technically Induced Change in the Temporal Organization of Radiological Work

Stephen R. Barley

HUMAN VIGILANCE'S tendency to trip over the apparently irrelevant contributes not only to the thrill of a good mystery and the power of hindsight but also to our periodic apprehension that the world is somehow shifting in directions we don't quite understand. As social scientists we profess to be well attuned to unanticipated consequences, to be steeled against irony. But in truth, our profession provides us no special immunity to surprise by the mundane. Like readers of mysteries riveted on dramatic turns of events, we often find ourselves outflanked by the subtle datum. As a case in point, consider research on the social ramifications of new technology.

Investigations of technologically driven social change in the workplace typically center on one of three topics. The majority grapple with economic issues—in particular, the promise of productivity and the fear of unemployment. In recent years, for instance, the "computerization of work" has nearly become a slogan for debate on the inevitability of occupational shifts, technical obsolescence, and the loss of work: the very topics that garnered so much attention in the 1950's, when, during another recession, the specter of automation first materialized in the public's eye.[1] The second and perhaps equally prominent line of inquiry examines the link between technical change and alienation. Since World War II, social scientists of various political persuasions have repeatedly sought to explicate Marx's contention that industrial modes of production alienate workers by showing how particular technologies shape social relations, structures of control, and perceptions of powerlessness, meaninglessness, and self-estrangement.[2] Although they use different vocabularies and offer contrasting theoretical visions, these same concerns remain pivotal today among those who write about deskilling and the quality of work in the wake of the "micro-electronics revolution."[3] The third cluster consists of studies that treat technology as a determinant of organizational structure. From this perspective, researchers usually seek to correlate attributes of technology with such contours of organizational form as horizontal and vertical differentiation or centralization and decentralization.[4]

This triad of economic, existential, and structural upshots arguably accounts for the most powerful and dramatic changes that technologies can induce in a workplace. Consequently, as matters of investigation, their importance is beyond dispute. Yet, while fully recognizing how critical these issues are for understanding technical change, we may nevertheless contemplate their hidden liability. By focusing continually on employment, alienation, and formal organization we may have unwittingly truncated the number of social parameters we hold responsive to technical change.

In the existing literature on technology and work, potential registers of social change are nearly exhausted once one considers the division of labor, routinization, distributions of authority and control, levels of staffing, and the nature of tasks and skill. But technology may alter more mundane characteristics of work as severely, or even more severely, than it affects these obvious parameters that so readily attract our attention. By looking exclusively at the overt characteristics of a workplace, we neglect forces that operate in the background to lend coherence to social organization. One of the most significant but subtle, and therefore easily forgotten, forces for order at work is the temporal structure of daily events.

TECHNOLOGY AND TEMPORAL ORDER

Aside from the allocation of activities to space, nothing anchors patterns of work more securely than the timing of action. Every work world carries the brand of a "socio-temporal order" (Zerubavel, 1981, p. 1), a structure mapped by sequences, durations, temporal locations, and rates of recurrence. Although sometimes loosely tied to physical or biological timetables, socio-temporal orders carry the force of an objective presence even when they are divorced from nature. In fact, one of the most potent techniques we humans have for turning culturally arbitrary behavior into social fact consists of our tendency to treat even self-imposed temporal boundaries as inviolable external constraints. Cycles, rhythms, beginnings, endings, and transition points not only support a social structure's aura of objectivity and predictability but also aid us in defining our roles, our obligations, and the tenor of our relationships. When faced with an ambiguous situation, one of the first and most important acts of sensemaking involves constructing a timetable of expected events and required behaviors (Roth, 1963; Van Maanen, 1977). The temporal order of the workplace therefore serves simultaneously as a template for organizing behavior as well as an interpretive framework for rendering action in the setting meaningful.

Despite the fact that temporal orders are necessary conditions for coordinated activity and even though temporal maps are elementary for fashioning shared social worlds, few social scientists have explicitly chosen to study time. Like the people they investigate, social scientists in general, and sociologists of organizations and occupations in particular, usually take time for granted. Against this void Zerubavel's (1979, 1981) detailed studies of temporal organization in hospitals, Roth's (1963) classic treatise on how doctors and patients negotiate the phases of a tubercular patient's hospital career, and Hall's (1959) sweeping analysis of temporal frames as hidden dimensions of intercultural conflict stand out as exceptions.

Given the sparse attention paid by sociologists to the broad topic of time in social life, it is hardly surprising that even fewer researchers have considered the more narrow question of how technologies might alter the temporal order of a workplace. A number of investigators have recognized that assembly lines dictate the pace of work (Walker and Guest, 1952; Chinoy, 1955; Blauner, 1964) and others have observed that automation almost inevitably leads to shift work (Mann and Hoffman, 1960). More specifically, Cottrell (1939) wrote of how a locomotive engineer's social existence revolves almost exclusively around the schedules by which trains ran. But beyond these and other scattered observations, there exist few systematic analyses of how technologies orient the temporal context of work. Even the early socio-technical theorists who repeatedly claimed that technologies alter spatial and temporal boundaries neither articulated a framework for analyzing such change nor offered an extended example of how temporal change might occur in an actual work setting (see Trist and Bamforth, 1951; Miller, 1959).

However, for those who would assemble a theory of how technologies shift temporal orders and thereby alter work milieus, Zerubavel's paired notions of "temporal symmetry" and "temporal asymmetery" offer a sturdy, if preliminary, analytic scaffold. Although Zerubavel developed the concepts to account for the rhythm of a modern hospital (1979, pp. 60–82) and the temporal organization of medieval Benedictine monasteries (1981, pp. 64–69), he proposed the notions as cornerstones for a general sociological treatment of time. Both concepts presuppose the existence of two or more actors (or groups) who may or may not live in similar temporal worlds. Temporal symmetry refers to situations where actors subscribe to "one particular pattern of temporal conditions" (Zerubavel, 1981, p. 64), where the rhythms of people's lives "duplicate" each other (Zerubavel, 1979, p. 60). In contrast, temporal asymmetry occurs when actors exist conterminously but subscribe to alternate temporal arrangements. The idea that people inhabit differently timed social worlds draws attention to how temporal orders support patterns of social integration and differentiation and thereby influence the dynamics of cohesion and conflict.

Zerubavel (1981, p. 65; 1979, p. 62) suggests that temporal symmetry undergirds mechanical solidarity. When people perceive their lives to flow in parallel, when they experience the same sequences, durations, temporal locations, and rates of recurrence of events, they are more likely to believe they share a set of circumstances and, on that basis, develop a sense of commonality. Perceptions of commonality are further enhanced when the temporal organization of a group's life inscribes boundaries that set the members of the group apart from others in their immediate surroundings. The minutely defined and rigidly scheduled duties of the Benedictine monk cadenced a temporal order greatly at odds with patterns of life outside the monastery and were, in part, consciously contrived to encourage communal identification within.

Unlike medieval monasteries or traditional societies, however, most contemporary workplaces are complex organizations populated by multiple groups that operate with different temporal frameworks.[5] This diversity in temporal orientation reinforces social differentiation and poses problems of integration more or less absent under conditions of temporal symmetry and mechanical solidarity. While shared temporal orders continue to bolster cohesiveness and perceptions of commonality among the members of particular groups, if the work requires members of different groups to interact, a person's daily experience is likely to be shadowed by temporal asymmetries. To achieve organic solidarity, the form of integration characteristic of highly differentiated social systems, different temporal structures must be made to mesh at critical points. Zerubavel argues that the modern institutions of planning, scheduling, and the precise measurement of time are all strategies for encouraging "temporal complementarity" among temporally asymmetric worlds. Regardless of such coordinating devices, however, all else being equal, temporally asymmetric interactions are still more likely to engender conflict than are relations among temporally symmetric actors.

To understand the possible links between technical change, temporal symmetry and asymmetry, and shifts in the organization of work, it is useful to distinguish between the *structural* and *interpretive* aspects of a temporal order. Structural attributes are perhaps best conceived as those external aspects of a temporal world that can be described more or less reliably by an independent observer.[6] Four structural parameters are particularly important for characterizing the external contours of a temporal system: the *sequence* in which events typically occur; the *duration* of these events, or how long each usually lasts; when events usually occur, or their *temporal location;* and finally, the *rates* at which particular events recur. Since temporal symmetry and asymmetry are essentially structural conditions, the temporal worlds of two actors may be defined as symmetrical or asymmetrical by comparing the sequences, durations, temporal locations, and rates of recurrence of events that each experiences over the course of a day. When all parameters are identical, the two temporal orders may be said to be symmetrical. On the other hand, as their structural parameters diverge the two become increasingly asymmetric.

People employ these structural parameters of temporal order to make sense of events that occur in the course of their work. Interpretations so cast compose the internal parameters of a temporal world, parameters that are not as immediately obvious to a casual observer. By evaluating events against a shared scheme of expected sequences, durations, temporal locations, and rates of recurrence, people judge whether they are bored, whether something is wrong, whether they have done a good job, or whether others have acted responsibly. Such interpretations not only enable us to lend meaning to events in our work worlds; they lead us to form opinions and make pronouncements about the behavior of persons operating in alternate temporal systems. It would appear, then, that structural asymmetries are necessary but insufficient conditions for social conflict. In addition, asymmetries must be interpreted in ways that justify contention.

Having distinguished between the structural and interpretive aspects of a socio-temporal order, we may sketch the rudiments of how new technologies may alter the social organization of a workplace by shifting the timing of events. Technologies are initially likely to affect the temporal organization of work by catalyzing change in the structural parameters of one or more temporal worlds. These structural changes, in turn, alter the balance of temporal symmetries and asymmetries. The restructured temporal parameters subsequently begin to influence the way participants interpret their tasks, roles, and relationships. Depending on the direction of change (toward symmetry or asymmetry), shifting structures and interpretations may warrant greater cohesion or greater conflict among the actors who populate the work setting.

To illustrate the foregoing discussion, consider the temporal changes new technologies have brought to one specific work setting: radiology departments in community hospitals. The following data were collected during an ethnographic study of two radiology departments, dubbed Suburban and Urban, during the year that each began to operate its first whole-body CT scanner.[7]

THE TECHNICAL AND SOCIAL ORGANIZATION OF RADIOLOGICAL WORK

Unlike most medical specialties that evolved as physicians began to concentrate on particular disease processes, organ systems, and patient populations, radiology's development has always been inextricably bound to technical innovation. The initial use of radiographic images of medical diagnosis followed shortly on the heels of Wilhelm Roentgen's discovery of x-rays in 1895. Since then, radiologists have continually collaborated with physicists and engineers to refine equipment and techniques for creating images of the body's internal anatomy (Dewing, 1962). In fact, only by gradually appropriating control over the medical use of radiographic technologies and by securing an exclusive license to interpret diagnostic images did radiology eventually attain the status of a medical specialty independent of its non-medical collaborators (Larkin, 1978, 1983).

Even though radiology's history floats on a steady stream of innovation, until the late 1960's most technical change came to radiology as incremental improvements to existing machines and techniques (Dewing, 1962). As recently as 15 years ago, the work of most radiology departments consisted almost entirely of the production and interpretation of radiographic and fluoroscopic studies, examinations that had been performed in one form or another since the 1920's.[8] The typical department of the 1960's was staffed by members of two distinct occupations: radiologists, physicians who specialize in the interpretation of diagnostic films, and radiological technologists, individuals with on-the-job training or an associate's degree in "radiological technology." During this period both occupations were internally undifferentiated: within each individuals could be distinguished from one another only in terms of personal experience and skill.

However, a rigid division of labor based on the segregation of productive and interpretive expertise separated the two occupations from each other and created a status hierarchy within radiology departments. Technologists managed patients during examinations and produced films for radiologists. Radiologists, in turn, extracted diagnostic information from the films and provided referring physicians with "readings." Although technologists were trained to run equipment and to recognize anatomy, they were not taught to interpret signs of pathology and, in point of fact, even after years of experience a "tech" might recognize few of the common ailments revealed by sets of x-rays (Barley, 1984). In contrast, however clumsy their attempts might appear when compared to the performance of a practiced technologist, most radiologists were able to operate x-ray machines and fluoroscopes. This pattern of expertise formed a hierarchy of authority in which the radiologists knew what the technologists knew, but in which the reverse did not hold.

In the mid-1960's a stream of new technologies began to infiltrate radiological work. Although often appended haphazardly to descriptions of technical change, as a succinct summary of how these machines cumulatively transformed the medical profession's ability to peer inside the human body, the term "technical revolution" may be warranted. The first technology to broaden the realm of diagnostic imaging was nuclear medicine, which entailed injecting radioisotopes into a patient's bloodstream and tracing their subsequent movement through the body with a "gamma camera," a device that records nuclear decay. Particular isotopes tend to congregate in specific organs and pathological entities that appear as "hot spots" on photographic film.

When rapid film changers were perfected in the late 1960's, special procedures also became common in the radiology departments of community hospitals. "Special procedures" is a cover term for a host of minor surgical techniques that require radiologists to implant catheters in a patient's body. Through the catheter an iodine dye is injected into either the vascular system or specific organs. As the dye courses through the body, a rapid series of x-rays is taken to assess conditions such as arterial occlusions or dilated bile ducts. With the advent of special procedures radiologists expanded their role to include the skills of minor surgery and thus ceased to be solely interpreters of films.

After picking up steam in the late 1960's, the pace of technical change in radiology quickened in the 1970's, when the computer's ability to transform data into pictures created new approaches to medical imaging. Of these computerized technologies, ultrasound and the CT (computerized tomography) scanner are the best known, but others, such as nuclear magnetic resonance (NMR) and positron emission tomography (PET), have recently made diagnostic debuts.[9] Ultrasound coupled the computer to a sonar device to create images of organs that were previously either obscure on x-rays or too susceptible to damage to be subjected to radiation. For example, with ultrasound radiologists can visualize the anatomy of a beating heart, detect pregnancy, and monitor fetal growth. The CT scanner was the first imaging device to record entire cross-sectional rather than longitudinal perspectives of internal anatomy. Since the resolution of a scanner's detailed pictures approaches the quality of an illustration in an anatomy text, its use has enabled radiologists to study organ systems and disease processes never before amenable to direct inspection—for example, the anatomy of the brain or the gradual spread of metastases through the abdomen. As computer-driven technologies, ultrasound and CT scanners both break radically with radiology's technical past. Each not only operates by principles dramatically different from those of traditional machines but, more importantly, creates a completely new semiotic system, a new language of diagnostic signs that radiologists must master. For the latter reason, the new technologies are often called "new modalities."

As they spread from medical centers where they were first developed to smaller community hospitals, the new technologies have gradually restructured the organization of radiology departments. The traditional occupational hierarchy based on a division of interpretive and productive labor has remained more or less intact, but the two formerly homogenous occupational groups have become increasingly differentiated. The ability to operate one of the newer technologies has gradually gained the status of a recognized subspeciality among radiological technologists. Whereas 20 years ago all technologists worked as "x-ray techs," today the typical radiology department employs individuals who call themselves "nuclear medicine techs," "specials techs," "sonographers," and "CT techs."

As the titles suggest, each suboccupation is primarily responsible for conducting examinations with its respective technology. Although most persons currently filling such roles are former x-ray techs who learned their new skills while on the job, many community colleges and vocational schools now offer formal programs in the new modalities that are distinct from, or appended to, the main curriculum of radiological technology. Sonography has progressed further toward a clear occupational identity than have the other suboccupations: it is now almost impossible to become a sonographer unless one has graduated from a program in ultrasound certified by the American College of Radiology. In most instances, technologists who operate the new technologies are better paid than their colleagues, the x-ray techs.

Eric Miller (1959) once proposed that differentiation of occupational roles by technology is almost always reinforced by spatial segregation. The topographies of most modern radiology departments support Miller's contention. Fluoroscopy, routine x-rays, and other long-established radiographic procedures are usually congregated in an area known in earlier times as the "Radiology" or "X-ray Department." While outsiders still refer to the same facilities as "X-ray," insiders more frequently call the area the "Main Department." The latter term suggests that members of a radiology department now divide their territory into core and periphery. In the Main Department work most of radiology's personnel: its secretaries, orderlies, clerks, and x-ray techs, as well as the department's radiologists when assigned to radiography and fluoroscopy. Unlike the core, the Main Department, the periphery has no specific identity and usually occupies no contiguous space. Instead, the periphery is composed of discrete locales each of which carries the name of the technology housed within its boundaries. More often than not, Ultrasound, CT, Nuclear Medicine, and Special Procedures are situated some distance from each other as well as from the Main Department. The separation of technologies in physical space and the new occupational roles combine to limit regular interaction among the various technologists the department employs and thereby reinforce perceptions of occupational difference (Barley, 1984).

The new modalities have yet to spawn formally recognized specialties among radiologists. Nevertheless, each technology has created its own experts so that radiologists tend to divide their labor informally.[10] To perform special procedures and interpret nuclear scans, sonograms, or CT scans requires extensive training and practice. Since most radiologists have few opportunities to learn new modalities before their department purchases its own device, at the time the department acquires a new technology it almost always also hires one or more radiologists who are already proficient in its use. The newcomer acts as an internal expert who takes responsibility for the technology's implementation and attempts to teach the modality's diagnostic language to his or her colleagues.

Although the department's longer-tenured radiologists usually share responsibility for the technology's daily operation and attempt to learn to read the images it creates, both they and the hospital's referring physicians are acutely aware of their novitiate status and recognize that it is the newest (and usually the youngest) member of the department who has the greatest expertise. Thus, while the new modalities have certainly broadened the radiologist's work, at the same time they have introduced a sub rosa form of specialization that inverts the radiologists' traditional status hierarchy, a ranking that presumed prestige to be a function of expertise measured in years of work experience (see Barley, 1986a).

By altering the tasks and roles of radiologists and technologists, each new modality has increased the number of work worlds in a radiology department. Among technologists, these worlds are experienced by different individuals who usually specialize in a particular technology. Radiologists, on the other hand, take part in several of the social contexts as they rotate daily or weekly among the technologies. Although the new work settings can be distinguished from each other along a number of critical dimensions, the social milieus surrounding the new technologies are, in general, more similar to each other than to the social context of the main department. For example, when compared with their counterparts in the X-ray Department, sonographers, CT techs, and specials techs have more variety in their work, are less closely supervised, are more likely to be able to interpret the films they produce, and have closer relations with referring physicians (Barley, 1984).

The most striking differences between the new and old technologies, however, pertain to the relationships that develop among the technologists and radiologists. In brief, such relationships in the newer technical areas are marked by greater equanimity and less conflict than are relations between the very same radiologists and the x-ray techs in the Main Department. The distinctions arise, in part, because new modalities intially level status differences based on distributions of knowledge (Barley, 1986b). However, other aspects of the social order surrounding the new technologies also contribute to qualitatively different forms of interaction. Important among these other parameters are the temporal arrangements that distinguish work in the Main Department from work in CT, Ultrasound, and Special Procedures.

THE TEMPORAL ORGANIZATION OF A MAIN DEPARTMENT

When assigned to radiography or fluoroscopy, radiologists at Urban and Suburban hospitals worked out of their own offices, which were adjacent to the Main Department. The only exception to this practice occurred during the morning at Urban hospital, when the radiologist assigned to fluoroscopy usually remained in the "fluoro rooms" until lunch, or until the morning's procedures were complete, whichever came first. When technologists, physicians, and administrators required a radiologist's assistance, they knew they could usually find one by either visiting the radiologist's office or calling on the radiologist's phone. Since radiologists made frequent trips to and from their offices as duties required, however, under the assumption that one could more easily corner a radiologist by presenting a physical presence, most individuals chose to summon or consult in person. Only referring physicians located outside the hospital routinely communicated with radiologists by phone.

The duties of radiologists assigned to the Main Departments were threefold. Besides routine x-rays, injections had to be administered, fluoroscopy conducted, or films reviewed at prescribed points in the flow of all radiographic procedures. Since the techs were prohibited from executing these tasks, whenever techs reached such a juncture in the course of a fluoroscopic or radiographic exam they had to seek the assistance of the radiologist on duty. Besides attending to these junctures in the flow of examinations, radiologists in the Main Department were also responsible for consulting with physicians and colleagues who sought verbal interpretations of films or advice on the efficacy of particular procedures. Finally, when not consulting or attending to a particular study, the radiologists reviewed and dictated readings on the seemingly endless sets of films that were produced over the course of a day in the Main Department.

The sequence, duration, temporal location, and rate of recurrence of these various activities defined the tempo of radiologists' work. To best gain an appreciation for the temporal organization of a radiologist's experience, examine the flow of events in Figure 4-1, which displays a five-hour segment in the work day of a radiologist assigned to the Main Department at Suburban. Although the chart maps occurrences in the life of one radiologist on one day, the structure of the depicted events typifies the rhythm of radiologist's work in both main departments. Discounting the angiogram (a special procedure) he performed in the morning as a favor for another radiologist (9:43 A.M. to 11:40 A.M.) and the hour and six minutes he spent at lunch (12:31 P.M. to 1:37 P.M.), the radiologist engaged in 53

Figure 4-1
The Course of a Typical Day of a Radiologist Assigned to a Main Department
(Suburban Hospital)

Beginning	*Duration (mins.)*	*Event/Action*
9:35 A.M.	4	Reads and dictates a study in his office.
9:39	1	Clerk enters and removes studies previously read. Informs radiologist that the study he requested has been mounted on microfilm machine.
9:40	2	Reads and dictates next study.
9:42	0.2	Specials tech summons radiologist to angio suite by telephone.
9:42	0.8	Finishes dictating study.
9:43	81	Performs a carotid angiogram in the angio suite.
11:04	3	Speaks with family of patient who had angiogram.
11:07	2	Returns to office. Reads and dictates another study.
11:09	0.5	Reads and dictates next study.
	0.5	Reads and dictates next study.
11:10	2	Reads and dictates next study.
11:12	2	Technologist enters with scouts from an IVP and summons radiologist to do injection.
11:14	1	Technologist leaves. Radiologist finishes dictating study.
11:15	7	Injects the IVP patient.
11:22	2	Begins to wash hands after IVP, but returns to exam room when patient begins to make odd noises.
11:24	2	Consults with a colleague about an ultrasound in colleague's office.

Figure 4-1 *(Continued)*
The Course of a Typical Day of a Radiologist Assigned to a Main Department
(Suburban Hospital)

Beginning	*Duration (mins.)*	*Event/Action*
11:26	2	Reviews an old study mounted on the microfilm reader.
11:28	2	Returns to own office. Consults with a urologist who is waiting there with films in hand.
11:30	1	Urologist leaves. Technologist arrives with IVP films.
11:31	1	Technologist arrives to summon radiologist to a barium enema.
11:32	2	Conducts the barium enema.
11:34	1	Consulted by a colleague about a knee film.
11:35	2	Returns to office and finds a urologist there. Consults with urologist about films urologist has hung on the lightbox.
11:37	3	Reviews a second study for urologist.
11:41	1	Urologist leaves. Reads and dictates next study.
11:42	1	Reads and dictates next study.
11:43	3	Notices a mistake on earlier reading. Reads study second time and redictates.
11:46	2	Consulted by a physician.
11:48	1	Physician leaves. Reads and dictates next study.
11:49	1	Reads and dictates next study.
11:50	2	Technologist enters with follow-up IVP films.
11:52	6	Tech leaves. Resumes reading study began at 11:49.

Figure 4-1 *(Continued)*
The Course of a Typical Day of a Radiologist Assigned to a Main Department (Suburban Hospital)

Beginning	*Duration (mins.)*	*Event/Action*
11:58	3	Technologist enters with films from a barium enema.
12:01 P.M.	1	Reads and dictates next study.
12:02	3	Two techs enter simultaneously, one with a post-void film from an IVP, one with a question about a problematic mammogram.
12:05	2	Resumes reading study begun at 12:01.
12:07	1	Reads and dicates next study.
12:08	6	Telephone call from colleague.
12:14	17	Reads and dictates "whole body" series.
12:31	66	Lunch
1:37	7	Cornered by a sonographer in hallway. Conducts a cardiac ultrasound.
1:44	15	Leaves ultrasound and is immediately summoned by a technologist to inject a young child having an IVP.
1:59	7	Provides wet readings for 4 technologists who have queued up outside his office during the IVP.
2:08	3	Consults with a colleague about an ultrasound study.
2:11	3	Summoned by technologist to a mammogram.
2:13	5	Summoned by a sonographer to review a cardiac ultrasound.
2:18	1	Returns to office. Technologist enters with films from an IVP.
2:19	2	Speaks with child's mother, who has requested information.

Figure 4-1 *(Continued)*
The Course of a Typical Day of a Radiologist Assigned to a Main Department (Suburban Hospital)

Beginning	*Duration (mins.)*	*Event/Action*
2:21	10	Consults with a physician.
2:31	3	Reads and dictates next study.
2:33	4	Reads and dictates next study.
2:38	1	Reads and dictates next study.
2:39	1	Technologist enters with films and summons him to a mammogram.
2:40	2	Takes tape of his dictations to the transcriptionists.
2:42	2	Attends mammogram.

SUMMARY (less angiogram and lunch):

Total time:	162	minutes (2.7 hours)
Number of activities:	53	activities
Activities per hour:	19.6	activities/hour
Mean duration of activities:	3	minutes
Median duration of activities:	2	minutes
Interruptions:		
by technologists:	14	interruptions
by colleagues:	4	interruptions
by referring physicians:	4	interruptions
Interruptions per hour:	8.1	interruptions/hour

separate tasks or encounters with a mean and median duration of three and two minutes, respectively. He was interrupted fourteen times by technologists and clerks, four times by colleagues, and four times by referring physicians for an average of approximately eight interruptions per hour.[11]

As the example portrays rather vividly, the flow of a radiologist's day in the Main Department was parsed by a large number of discrete events, all of relatively short duration. Moreover, the precise sequence of events was essentially unpredictable and the location of any event in the passage of time could only be vaguely estimated. Radiologists expected to be consulted by physicians and colleagues, but they never knew when such consultations would occur. Because specific types of exams were scheduled for certain periods of the day, radiologists could predict that technologists were more likely to summon them to intravenous pyelograms in the morning and to mammograms in the afternoon, but the precise timing of these summons could not be foreseen. Aside from the ever present pile of unread films, it was also impossible for a radiologist to estimate from the vantage point of his office the rate at which any activity would recur.

Thus, the temporal organization of a radiologist's work was such that at any moment he could be drawn for a brief span of time into the work worlds of any of a large number of individuals. It was as if the radiologists existed at the nexus of a number of trains of action that ran on unpredictable schedules and made unanticipated stops. Since the temporal boundaries of the day were extremely fluid and the day's punctuation into segments was largely out of their control, radiologists usually accommodated to the temporal structure by treating the timing of events flexibly. All tasks would occur at their due moment. Since the nature of the work precluded the radiologists from dealing with one event at a time, they came to prefer multiple lines of simultaneous action. Most radiologists at both hospitals admitted that they found the hectic pace exhilarating and that other tempos bored them.

In comparison to the disorderly and unpredictable temporal world of the radiologist, the rhythm of an x-ray tech's work was more certain and well behaved. Each day both departments posted their x-ray techs to one or more duty stations, where they performed specific types of procedures for specified periods of time. A technologist might, for example, conduct intravenous pyelograms or fluoroscopy in the morning and routine x-rays or mammograms in the afternoon. The cyclic beginnings and endings of a series of such procedures segmented the flow of a technologist's work into discrete units.

Whereas numerous brief encounters marked the character of a radiologist's day, the examinations that parsed the technologist's work day ranged in duration from an average of 6 minutes for routine x-rays to an average of 56 minutes for intravenous pyelograms (IVPs).[12] In addition to being of comparatively longer duration, protocols and appointment schedules programmed the sequence, temporal location, and rate of recurrence of most of a technologist's activities. When assigned to work any area of the Main Department aside from routine x-rays, technologists executed the same examination or one of a small set of examinations for the duration of the duty assignment. Between each exam's beginning and ending flowed a series of well-rehearsed acts that unfolded linearly in lock-step fashion. This chain of tasks, identical across all repetitions of the procedure, was broken only when technical glitches or the rare medical emergency interrupted the normal flow of work.[13] Since most procedures were scheduled ahead of time, x-ray technologists could estimate the temporal location of each activity cycle by consulting appointment sheets.

Ironically, only when performing "routine" x-rays did technologists' work become less predictable and well structured. Both departments scheduled few routines in advance. Rather, patients who were to have "plain films" arrived randomly from the floors or through the outpatient clinic. Since the term "routine" covered a large number of different studies, technologists assigned to these duties could not predict the exact nature, sequence, or temporal location of their next activity. It was because of this variability that most x-ray techs preferred to work "routines." But even when doing routine x-rays, once technologists began a particular study, such as a chest x-ray or a skull series, their actions unfolded by a script as rigid as the one that dictated the moves of technologists doing IVPs or fluoroscopy.

Although "routine x-rays" brought technologists a measure of temporal variety, to conclude that a loose temporal structure usually inscribed the technologist's work day would be mistaken. On a given day most of Urban's and Suburban's x-ray techs administered barium enemas, upper gastrointestinals, IVPs, mammograms, and other highly structured, scheduled exams. Even when they did work "routines," technologists often spent at least part of the day performing other procedures. Consequently, from the technologists' point of view, the tempo of most work was exceedingly predictable, a perception underscored not only by the techs' preference for "routines" but also by their half-serious jests about fluoroscopy, mammograms, and IVPs being "factory work." According to most technologists, the only sure uncertainty that plagued the temporal orderliness of their day was the need to engage a radiologist.

Had there been no necessity of bringing radiologists into the flow of a procedure to inject a patient, to conduct fluoroscopy, or to review preliminary films, the asymmetry between the temporal structure of the technologists' and the radiologists' work would have remained an unimportant curiosity with few ramifications for social relations in the department. Routine x-rays, for example, engendered few hostilities over the timing of action precisely because technologists could almost always complete these exams without seeking a radiologist's assistance or counsel. When doing IVPs, fluoroscopy, mammograms, or any of a host of other procedures known as "subspecials," however, technologists could not continue the study past a predetermined point without a radiologist's involvement. It was the need to summon a radiologist that made technologists aware of the potential discrepancy between their structured, linear world and the temporally unpredictable world of the radiologist.

To locate a radiologist, a technologist often had to search several offices and ask other technologists about the radiologist's last known whereabouts. Even after the tech found a radiologist, there was no guarantee that he would be immediately available. At the time of the tech's arrival, the radiologist could be talking on the telephone, discussing a film with a physician, consulting a colleague, or about to assist with another examination. In each instance the technologist would have to wait. But even if the technologist successfully engaged the radiologist's attention, he or she still had no firm claim on the radiologist's time. The radiologist could always be diverted by a number of events, including a telephone call, a consultation, or even another technologist with a request that the radiologist deemed more important. On numerous occasions, I witnessed a technologist summon a radiologist to an IVP or fluoroscopic exam only to discover, several minutes later, that the radiologist had forgotten to follow the technologist to the examination room. In such instances the tech grudgingly delivered a second summons after granting the radiologist a period of grace to show up on his own.

When an initial summons failed to mesh the two temporal flows successfully, technologists relied on their own notions of temporal propriety to account for the ruptured work flow in terms that held the radiologist capable and worthy of at least passing condemnation. X-ray techs consistently complained that the most irritating aspect of their work was that "you can never find a radiologist when you want one." The techs explained this difficulty in a number of ways. Suburban's x-ray techs used such situations to justify their claim that the department needed to hire an additional radiologist, a recommendation occasionally used to needle radiologists who neglected a technologist's summons. More importantly, at both sites, the x-ray techs proclaimed in often-repeated asides that the difficulty of securing a radiologist attested to the radiologists' callousness: "They think nothing of keeping a patient waiting twenty minutes." Even in Urban's "fluoro area," where radiologists remained nearby for most of the morning, the techs became indignant whenever a radiologist left the area for reasons they thought inappropriate.

From the technologists' point of view, the radiologists' adaptive, fluid approach to an unpredictably structured temporal world appeared, at times, to be little more than blatant disregard for the prescribed order of events that the techs were obligated to uphold. Their moral indignation was heightened by the fact that patients also experienced examinations as a linear flow. If an obvious delay occurred between the phases of an exam, patients were more likely to complain to the technologist than to the radiologist, who typically interacted with the patient only for a few brief minutes. Thus, the technologists argued that they were unfairly, but ultimately, held accountable for a radiologists' delays.

Radiologists, on the other hand, perceived technologists to be unreasonably demanding and rigid in their expectations. To the radiologists, technologists sometimes appeared to presume that their current examination was the most pressing concern the radiologist faced. The radiologists, therefore, contended that technologists could not accurately assess the relative priority of events. Radiologists might respond to a technologist's pressure or castigation by claiming that the technologist was being unduly "hyper" or "bitchy." Because each party's interpretation of the other's action was framed against an alternate temporal framework, an ever-present potential for open contention and hostility undulated silently beneath the act of summoning in both main departments.

THE TEMPORAL ORGANIZATION OF THE NEW TECHNOLOGIES

Specials techs, CT techs, and sonographers complained far less frequently about not being able to find a radiologist, even though they, like the x-ray technologists, summoned radiologists at some point during almost every exam. One might, at first, hypothesize that the difference reflects alterations in the temporal organization of technologists' work. However, the structures of the technologists' temporal worlds differed little from those found in the Main Department. Beginnings and endings of examinations parsed the CT tech's, the specials tech's, and the sonographer's work day into discrete segments. Scheduled appointments set the sequence and temporal location of every special procedure, CT scan, and ultrasound performed in both hospitals. Technologists who operated new modalities could estimate the rate at which particular activities would recur by consulting the appointment book that they themselves often kept. Like the studies conducted in the Main Department, each CT scan, ultrasound, and special procedure consisted of a sequence of prescribed actions reiterated whenever technologists performed a particular type of study. Finally, in comparison to the events that marked the work day of a radiologist in the Main Department, the activities of a technologist running a new modality were of lengthy duration. The average CT scan spanned 48 minutes, the average special procedures lasted 90 minutes, and the average ultrasound ran for 16 minutes.[14] Thus, when working the newer modalities, technologists experienced temporal worlds that were as linear and as cyclically predictable as those experienced by the x-ray tech in the Main Department. To explain why summoning a radiologist engendered little or no conflict in the context of the newer technologies requires examining the temporal structure of radiologists' work.

When assigned to CT duty, the radiologists at both

hospitals worked out of an office located in the CT area. On these days, Urban's radiologists attended only to scanner work and consulted mainly with physicians and colleagues interested in reviewing or ordering scans. Although Suburban's radiologists combined CT and ultrasound duty, together the two technologies produced fewer studies than did the Main Department. And, like their counterparts at Urban, when assigned to the CT Department, radiologists at Suburban consulted more or less exclusively with physicians who ordered scans or ultrasounds. By limiting the scope of the radiologist's responsibilities and by distancing the radiologist from other technologies, CT duty reduced the variety and number of the radiologist's activities while increasing the duration of the events that marked the flow of his day.

Figures 4-2 and 4-3, respectively, display randomly selected examples of a radiologist's day in the CT Departments at Suburban and Urban. Note that the duration of a radiologist's average activity is greater than that observed in the Main Department. In these examples the average activity lasted approximately six and seven minutes respectively, while in the previous example of work in the Main Department (Figure 4-1), the average act spanned three minutes. It follows that the pace of the radiologist's day was slower when assigned to a CT. In Figure 4-1, the radiologist in the Main Department engaged in 19.6 activities per hour. In the two CT examples, this pace was approximately halved: on the days depicted, the first radiologist experienced 10.4 and the second radiologist 8.3 activities per hour.

While activity counts intimate that the tempo of radiologists' work differed in the CT area and the Main Department, they gloss an extremely important qualitative difference in the temporality of the two settings. Situated in the CT area with their responsibilities limited primarily to scanner work, the radiologists found their days more

Figure 4-2
The Course of a Typical Day of a Radiologist Assigned to the CT Department (Suburban Hospital)

Beginning	*Duration (mins.)*	*Event/Action*
11:30 A.M.	14	Studies an abdominal scan hanging on lightbox in CT office.
11:44	6	Goes to main department to consult with a colleague about the abdominal scan.
11:50	10	Returns to CT office and dictates a reading on the abdominal scan.
12:00 noon	5	Summoned by a CT tech to console room to review a scan.
12:05 P.M.	1	Returns to CT office. Finishes dictation.
12:06	3	Hangs and reviews next scan.
12:09	9	Goes to main department to consult a colleague about the scan.
12:18	5	Returns to CT office. Calls the patient's physician for information.
12:23	5	Finishes telephone call. Dictates a reading on the scan.
12:28	5	Visits console room and speaks with technologists about course of current scan.
12:33	6	Administrator enters console room. Discusses a radiation exposure analysis with administrator.
12:39	5	A physician enters the console room. Administrator leaves. Physician consults radiologist about scan in progress.
12:44	2	First physician leaves. Second Physician arrives for consultation. The two go to CT office.
12:46	13	Physician leaves. Radiologist goes to the console room to speak with technologists.

Figure 4-2 *(Continued)*
The Course of a Typical Day of a Radiologist Assigned to the CT Department (Suburban Hospital)

Beginning	*Duration (mins.)*	*Event/Action*
12:59	8	Returns to CT office and reviews scans.
1:07	0.2	Summoned by sonographer to an ultrasound exam by telephone.
	5.8	Leaves to review the ultrasound.
1:12	2	Returns to office. Dictates reading on the ultrasound.
1:14	1	A colleague calls on the phone to suggest that the radiologist go to lunch and that he will cover CT in the radiologist's absence.
1:15	3	Goes to the console room to speak with the technologists.
1:18	58	Lunch
2:16	12	Returns to CT office. Calls a colleague on phone to consult about a scan.
2:28	6	Completes call. Technologist enters with previous scan on the present patient. Radiologist reviews old scan.
2:34	9	Goes to main department to consult with a colleague about a scan.
2:43	2	Stops in hall on way back to CT to consult with another colleague about the scan.
2:45	4	Returns to CT office. Dictates reading on scan.

Summary (less lunch):

Total time:	141	minutes (2.4 hours)
Number of activities:	25	activities
Activities per hour:	10.4	activities/hour
Mean duration of activities:	6	minutes
Median duration of activities:	5	minutes

Figure 4-2 *(Continued)*
The Course of a Typical Day of a Radiologist Assigned to the CT Department (Suburban Hospital)

Beginning	*Duration (mins.)*	*Event/Action*	
Interruptions:			
by technologists:		3	interruptions
by colleagues:		1	interruptions
by referring physicians:		3	interruptions
Interruptions per hour:		2.9	interruptions/hour

closely tied to the scanner's schedule, the very round of activity that shaped the temporal organization of the CT techs' experience. Whereas the segmentation of radiologists' work in the Main Department was never ordained in advance, the beginning and endings of CT scans lent to the radiologist's day a cyclical structure open to anticipation. By monitoring the day's schedule and the progress of the technologists' work, a radiologist could estimate more or less accurately the sequence, temporal location, and rate of recurrence of his primary duties. Although radiologists were still unable to foresee when physicians or colleagues would initiate consultations, these uncertainties were now projected on a background of predictability rather than on a canvas of random action.

In Figure 4-1, the example from the Main Department, the radiologist spent little time with any person or procedure. Not only do the chronicles from CT duty show radiologists constantly checking on the course of a scan; in both examples an appreciable amount of the radiologist's time (26% at Suburban and 68% at Urban) was spent at the console or in the gantry room, where the radiologist and the technologists simultaneously experienced the same chain of events. These alterations in the structure of the radiologist's day increased the degree of symmetry between

Figure 4-3
The Course of a Typical Day of a Radiologist Assigned to the CT Department
(Urban Hospital)

Beginning	*Duration (mins.)*	*Event/Action*
9:52 A.M.	4	Reviews a scan and dictates reading in CT office.
9:56	10	Goes to console room. Reviews a scan on monitor while techs prepare next patient.
10:06	2	Enters gantry room. Starts contrast at technologist's request.
10:08	2	Returns to console room. Talks to technologists.
10:10	3	Summoned by a technologist to inject patient in head scanner.
10:13	31	Returns to body scanner's console room. Observes a scan with technologists.
10:44	4	A physician enters console room. The two go to CT office, where physician consults radiologist about scan.
10:48	7	Physician leaves. Radiologist returns to console room.
10:55	2	Enters gantry room to start contrast.
10:57	8	Returns to console room.
11:05	2	Returns to CT office. Reviews a scan.
11:07	3	Returns to console room.
11:10	3	Summoned by a technologist to head scanner to review images.
11:13	18	Returns to body scanner console room and sits beside techs at console.
11:31	25	Returns to CT office. Reviews and dictates reading on the scan.
11:56	6	Goes to main department to consult with a colleague.

Figure 4-3 *(Continued)*
The Course of a Typical Day of a Radiologist Assigned to the CT Department (Urban Hospital)

Beginning	*Duration (mins.)*	*Event/Action*
12:02 P.M.	3	Returns to console room to check in with technologists.
12:05	5	Enters CT office. Reviews a scan.
12:10	3	A colleague enters to consult about scan.
12:13	1	Summoned by a technologist to inject a patient. Colleague leaves.

Summary:

Total time:	142	minutes (2.4 hours)
Number of activities:	20	activities
Activities per hour:	8.3	activities/hour
Mean duration of activities:	7	minutes
Median duration of activities:	3.5	minutes
Interruptions:		
by technologists:	3	interruptions
by colleagues:	1	interruptions
by referring physicians:	1	interruptions
Interruptions per hour:	2	interruptions/hour

the temporal organization of radiologists' and technologists' work. Consequently, when technologists reached a point in a scan where they needed the radiologist's assistance or counsel, they had very little difficulty engaging his attention. The radiologist was likely to be situated nearby, to have foreseen their summons, and to be able to disengage from his current activity to attend to the technologist's request.

Temporal symmetry between the work worlds of technologists and radiologists moved toward an even closer *isomorphism* in special procedures. Because they were minor surgeries, special procedures required a radiologist's constant presence once the exam actually began. Although specials technologists usually "prepped" patients before summoning the radiologist, from the time the radiologist arrived until the time the procedure was deemed complete, the radiologist and the specials tech literally worked side by side experiencing the same flow of events. Consequently, during special procedures the temporal worlds of every participant, including the patient, moved in parallel.

Even the initial summoning of a radiologist to a special procedure progressed more smoothly than similar summons delivered in the Main Department. Based on the preparation's progress, specials techs could estimate the time at which the procedure should begin. Using these estimates, they telephoned the radiologist in his office several minutes ahead of time to inform him that the patient was ready (as occurs in Figure 4-1). By the time the specials technologists were ready for the radiologist, he had almost always arrived. X-ray techs, by contrast, only sought radiologists in person and at that point in time when they actually needed assistance. Since special procedures were scheduled in advance, radiologists always knew the sequence and temporal location of their upcoming activities and could plan their participation accordingly. Moreover, few activities took precedence over a special procedure. If several specials were booked back to back, radiologists often remained in the specials area between exams, thereby eliminating even the necessity of the technologist's summons. In such instances, which were common in both hospitals, contention over the coordination of temporally asymmetrical action became moot since the two temporal worlds moved in unison.

In neither hospital did radiologists ever assume ultrasound duty as a sole responsibility. Before the scanners arrived, radiologists at both sites appended obligations for ultrasound to duty assignments in the Main Department. As mentioned earlier, after the scanner began operating,

Suburban's radiologists joined ultrasound and CT to form a new duty station. Urban's radiologists continued to link ultrasound to film reading in the Main Department until near the end of the study, when Radiology gained control of the hospital's formerly independent Nuclear Medicine Department.[15] Ultrasound and nuclear medicine duties were then combined. In each case, however, radiologists still stationed themselves some distance from the ultrasound area. Therefore, before releasing a patient, Suburban's sonographers regularly summoned a radiologist to review "real time" images while Urban's sonographers typically carried films or "hard copies" to a radiologist for scrutiny. Since the radiologist and the sonographer were not situated in the same physical space, their temporal worlds were not as symmetrical as those found in the CT department or special procedures. However, sonographers still experienced less frustration when attempting to engage a radiologist than did the x-ray technologists.

The frequency of contention arising from temporal asymmetry was lower in ultrasound for a combination of reasons. First, many radiologists placed higher priority on the sonographer's summons since they considered reviewing ultrasounds more interesting and challenging than their duties in the Main Department. Second, by the end of the study, radiologists at both hospitals were responsible for only one other modality in addition to ultrasound. Therefore, they were not as likely to be pulled simultaneously in as many directions as when assigned to the Main Department.[16] With fewer activities the tempo of radiologists' work seemed less hectic, if not more predictable. Consequently, they could respond in a more timely fashion than could the radiologist responsible for traditional procedures.[17]

Finally and perhaps most importantly, sonographers could circumvent the need to engage a radiologist when encounters promised undue delays. Unlike most radiological technologies, "real time" ultrasound equipment creates what may be likened to an ongoing motion picture of the area of the body under study. Although static photographs of images frozen on the machine's video monitor were the typical media for recording an ultrasound's results, it was possible to store an entire ultrasound procedure on video tape. When a radiologist could not attend to a procedure promptly, sonographers at both sites were allowed to tape the examination for the radiologist's later perusal.

Some sonographers employed a second method for allaying the frustration of meshing the flow of their temporal world with that of the radiologist. In sharp contrast to most radiological procedures, ultrasounds can not be produced unless the operator can interpret the ultrasound's images. The images guide the operator cybernetically to subsequent phases of the procedure by suggesting refined vantage points or hypotheses about what should be observed next. Therefore, unlike other technologists, sonographers had to recognize pathology in order to film it. Over a period of years, accomplished sonographers gained a reputation for competence among radiologists. Once a radiologist trusted a sonographer's competence he was likely to admit that his presence during an exam was little more than a formality unnecessary to the completion of an adequate exam.[18] Such perceptions provided sonographers with degrees of freedom to alter temporal constraints, opportunities that few other technologists enjoyed. For example, the most experienced sonographer at Urban routinely drew on his reputation to dismiss patients when radiologists were unable to review his films in a timely fashion. Although the practice was not formally sanctioned, radiologists never took issue with the sonographer's decision "to send" a patient without first seeking their counsel. Thus, ultrasound technology fostered skills that allowed sonographers to avoid conflict by *decoupling* their temporal world from the temporal world of the radiologist.

In sum, all new radiological technologies at Urban and Suburban, for somewhat different reasons, altered the temporal arrangements and probabilities of conflict experienced by radiologists and technologists working in the Main Departments. In each instance, the upshot was that at critical junctures it became less troublesome to integrate the radiologists' and technologists' temporal worlds. CT scanning and special procedures enhanced the temporal symmetry of radiologists' and technologists' work by restructuring the duration, sequence, temporal location, and rate of recurrence of events in the radiologists' day so that the radiologists' experience became more closely aligned with the flow of the technologists' work. The two originally asymmetric temporal frameworks moved toward what became, in the extreme case of a full docket of special procedures, temporal isomorphism. Ultrasound augmented any increase in symmetry by offering opportunities to abolish the possibility of conflict by decoupling temporal frames. At junctures where under other circumstances the temporal worlds of radiologist and sonographer might fail to mesh, sonographers drew on the technology's capacity and their own reputation for competence to avoid unpleasant and frustrating encounters entirely. Unlike the technologies in the Main Department, the new technologies enhanced the complementarity of temporal structures and thereby diffused interpretations that might warrant contention and conflict.

EPILOGUE: MONOCHRONIC AND POLYCHRONIC WORK CULTURES

The contrasting structural and interpretative aspects of the temporal orders that shaped the day-to-day experience of radiologists and technologists in the two Main Departments followed contours that parallel Edward Hall's (1959) distinction between "polychronic" and "monochronic" cultures. According to Hall, in a polychronic culture individuals place less value on temporal order, tend to accept events as they arise, and engage in multiple activities simultaneously. In contrast, people from monochronic cultures seek to structure activities and plan for events by allocating specific slots of time to each event's occurrence. Hall suggests that when individuals from polychronic and monochronic cultures are forced to interact, conflict and tension arise, a point he elaborates by discussing the temporal orientations of northern and southern Europeans:

> I have described two ways of handling time, monochronic and polychronic. Monochronic is characteristic of . . . peoples who compartmentalize time; they schedule one thing at a time and become disoriented if they have to deal with too many things at once. Polychronic people . . . tend to keep several operations going at once like jugglers. Therefore, the monochronic person often finds it easier to function if he can separate activities in space, whereas the polychronic person tends to collect activities. . . . Monochronic Northern Europeans find the constant interruptions of polychronic Southern Europeans almost unbearable because it seems that nothing ever gets done. Since order is not important to the Southern Europeans, the customer with the most "push" gets served first even though he may have been the last to enter. (Hall, 1969, p. 173)

Regardless of the veracity of his claim, the discomfort that Hall attributes to northern Europeans in the preceding excerpt appears analogous to the sentiments of x-ray technologists forced to operate on radiologists' time. The difficulty of summoning a radiologist violated the technologists' sense of order, a sense made more acute by the perception that patients also operated in a monochronic world. Much of the hostility that x-ray techs felt toward radiologists arose from the need to link their own monochronic work world to the polychronic world of the radiologist long enough to get the radiologist to perform an act that would preserve the examination's linear flow. Like attempting to engage a high-speed flywheel with a low-speed clutch, the tech's attempt to engage a radiologist often resulted in the social equivalent of a grinding noise.

In Hall's terms, special procedures and CT scanning can be seen to have altered the odds of cultural clashes between radiologists and technologists by increasing the monochronicity of the radiologists' world. Ultrasound, on the other hand, did little to change the temporal aspects of the radiologists' or technologists' work cultures, but rather created conditions where members of the two groups did not need to interact except under circumstances that violated neither party's assumptions about the rightful nature of temporal obligations. By joining Hall's notions of polychronic and monochronic cultures to the structural and interpretive aspects of temporal orders discussed in the body of the essay one may derive a more complete picture of how technologies changed the social organization of work in the two radiology departments under scrutiny.

Technological changes, such as the arrival of a new imaging modality in a radiology department, occasion a reallocation of activities or tasks. The verb "occasion" is critical, for although technologies influence the actual nature of work, the allocation of tasks to persons and persons to physical space are human choices made on grounds that are at least partially independent of technology's dictates. Regardless of their source, however, such allocations influence the temporal structure of a person's work day.

All else being equal, a large number of disparate tasks may sire the type of temporal structure that characterized radiologists' work in the Main Departments: an unpredictable sequences of events of short duration whose temporal location and rate of recurrence cannot be foreseen and therefore programmed. Under such circumstances, the members of an occupational group are likely to devise a constellation of interpretations that Hall calls a polychronic view of time. In contrast, activities so allocated to create cyclic sequence of events of longer duration with scheduled temporal locations and rates of recurrence may foster alternate interpretations that, when combined, yield the monochronic perspective. The first ramification of technical change for the social organization of work may therefore be posed: the creation of temporal subcultures rooted in the structure of the events typically experienced by members of an occupational group.

Additional ramifications may spring from the existence of different temporal subcultures if members of the various occupational groups are required to interact. If technologies occasion activities that sustain alternate temporal structures, then any required merging of actors who subscribe to different temporal orders will carry the possibility of conflict. On the other hand, if the temporal structures experienced by various occupational groups become more congruent after the arrival of a technology, then interaction may lead to less conflict than in previous times when subcultural proclivities were more disparate. Hence, new technologies may enhance or inhibit conflict by triggering changes in the structural allocation of events that, in turn, shift interpretive temporal frameworks. In the case of the two Radiology Departments, each new imaging modality nullified the temporal grounds for contention that characterized relationships surrounding the use of older radiological technologies.

Although I have tried, in this chapter, to trace one path by which technologies may alter the social organization of a workplace by shifting the socio-temporal orders that undergird occupational subcultures and interoccupational conflicts, I do not mean to imply that technologies cannot alter the timing of work in other important ways and therefore generate other forms of social change. For example, under certain circumstances, such as cheap off-hour CPU (central processing unit) time and restricted budgets, computers may transform what has traditionally been day work into night work, a metamorphosis whose social importance is likely to be especially well understood by faculty and students at numerous universities. My intent has been to show how technologies can set loose substantial changes in the social organization of work by tampering with crucial but overlooked parameters of social order—in this case, the timing of events.

If there is a moral to this analysis it is the following: we may prudently ask ourselves whether subtle and apparently superfluous parameters of social organization are not continually pulling end runs around our stalwart lines of traditional inquiry regarding the social implications of technical change. Perhaps we might better comprehend why we often sense that technologies are changing our worlds in directions we don't quite understand by paying less attention to the dramatic and more attention to the mundane. After all, despite our concern for the dramatic, most social worlds are rarely altered by cataclysm and most social change seems to creep up from behind.

NOTES

Acknowledgment: The research reported here was sponsored, in part, by a doctoral dissertation grant (HS 05004) from the National Center for Health Services Research.

1. Consider the words of Norbert Wiener, MIT's celebrated mathematician who coined the term "cybernetics" to refer to "control and communication in animals and machines": "Let us remember that the automatic machine, whatever we think of any feelings it may have or may not have, is the precise economic equivalent of slave labor. Any labor which competes with slave

labor must accept the economic conditions of slave labor. It is perfectly clear that this will produce an unemployment situation, in comparison to which the present recession and even the depression of the thirties will seem a pleasant joke" (1950, p. 169). Although Wiener wrote these words almost 35 years ago, they ring as if they were from a text written yesterday.

2. The study of technically induced alienation was the hallmark of American sociology of automation during the 1950's and 1960's. Key contributions to this genre include early investigations of the automobile industry (Walker and Guest, 1952; Chinoy, 1955; Faunce, 1958, 1965), as well as Mann and Hoffman's (1960) pioneering study of an automated power plant and Blauner's (1964) comparative analysis of the social implications of craft, machine tending, assembly line, and continuous process technologies. At approximately the same time, researchers at the Tavistock Institute in London inaugurated the British counterpart of the American sociology of automation, the study of socio-technical systems (see Trist and Bamforth, 1951; Trist and Murray, 1958; Rice, 1958; Herbst, 1962). Whereas the American literature focused on the social psychological response of the individual worker, socio-technical research centered on the work group.

3. Ideas found in the current literature on job design and the quality of work life can be traced directly to either the sociology of automation or the study of socio-technical systems (see Hackman and Oldham, 1981; Hackman and Lawler, 1971; Mumford and Weir, 1979). The deskilling literature, in contrast, is typically Marxist in its orientation and proposes that technologies alienate labor to the degree that they reflect the managerial class's desire for control (see Braverman, 1974; Noble, 1979, 1984; Shaiken, 1979).

4. Representative of organizational theory's treatment of technology are the following: Woodward (1958); Perrow (1967); Child (1972); Comstock and Scott (1977); Hage and Aiken (1969); Hickson et al. (1969); and Blau et al. (1976).

5. Lawrence and Lorsch (1967) recognized that functional groups within an organization, such as Sales, Research and Development, or Manufacturing, subscribe to distinctive time

frames. By and large, however, they restricted their consideration of temporal orientations to "time horizons," the relative distance into the future that members of each function consider within their purview.

6. I have chosen to speak of "external" and "internal" attributes of a temporal order rather than "objective" and "subjective" attributes since the latter suggests an ontological distinction that ultimately fails. While established sequences and durations are perhaps more easily observed than the meanings people attribute to temporal markers, they are no less the products of a socially constructed reality. Aside from social convention there is often no reason that events should occur in a certain sequence or last for specified periods of time. Similarly the distinction between "etic" and "emic" description does not capture the distinction I wish to make. More often than not, participants in a social world will perceive the structural aspects of a temporal order in precisely the same way as the observer perceives them. However, the structural, as opposed to the interpretive, attributes of a temporal order are more "accessible" to the observer. For this reason, I have chosen to call structure an external attribute and interpretation an internal attribute. The distinction between external and internal is similar to that made by Schein (1981) and Van Maanen et al. (1977) in their discussions of career lines.

7. The study was designed to compare the social organization of work surrounding all technologies used by the two departments and to document the evolution of the two scanner operations. At issue was the question of whether radiological technologies alter the roles and role relationships of radiologists and technologists and thereby change the structure of a radiology department. The study and its design are more fully explicated in Barley 1984, 1986a, and 1986b.

8. Radiography refers to the production of static images on photographic film. Included under this rubric are the numerous "routine x-rays" or "plain films" familiar to most people, as well as a number of other types of studies including tomography and several contrast-enhanced examinations such as the intravenous pyelogram (IVP). In fluoroscopy, a continuous stream of x-rays is projected onto a device known as an "intensifier." As electrons

strike the phosphorescent material of the intensifier, electrical impulses are fed to a television monitor that displays a constant, or "real-time," image of the area of the body under study. Thus, fluoroscopy allows the study of motion. Representative fluoroscopic exams include the barium enema and the barium swallow.

9. At present, NMR and PET are found only in large medical centers and are still considered experimental. NMR uses two electromagnetic fields to momentarily realign the axes of the nuclei of specific types of atoms. When one field is broken, the nuclei bounce back to their original positions, emitting energy that can be detected and transformed into images by a computer. Since NMR can be tuned to various types of atoms, some physicians hope the device will allow them to study disease processes rather than structure. PET scanning uses a computer to construct cross-sectional images of the body by measuring the decay of radioisotopes with very short half-lives. Since different isotopes have affinities for different organ systems, once again the promise is an opportunity to investigate actual pathological processes rather than simply image the structural changes brought about by pathology.

10. In large medical centers where research is conducted, radiological specialties associated with particular modalities have evolved. There, a radiologist who works with a CT scanner is, for example, unlikely to have duties in the Main Department or to perform special procedures. However, smaller community hospitals with limited financial and human resources can rarely afford such specialization.

11. All radiologists in both Urban and Suburban hospital were male. For this reason I use the masculine form of the third person pronoun throughout the remainder of the essay when referring to radiologists.

12. For all examinations observed in their entirety, I recorded beginning and ending times so that I might calculate mean durations for types of procedures. The following list records the mean duration (in minutes), the number of observations, the standard deviation of the mean, and the minimum and maximum duration of the actual procedures I timed in the Main Departments:

Procedure	*Mean*	*N*	*St. Dev.*	*Min.*	*Max.*
Routines	6	11	3.8	1	13
IVPs	56	28	21.6	27	127
Barium enemas	43	48	13.7	18	76
Upper GIs	15	41	7.7	7	51

13. Emergencies in the Main Department included such events as a patient suffering arrythmia on the examination table, gastrointestinal bleeding, or an epileptic fit in the hallway. Milder maladies, such as vomiting, were treated as everyday occurrences, as glitches that could slow a procedure but that were unlikely to bring the procedure to an end. During the course of the study no patient died while visiting the two Radiology Departments. However, older technologists could usually recite stories about patients who "coded" on the x-ray table.

14. Descriptive statistics on the duration of the CT scans, special procedures, and ultrasound examinations I observed are as follows (all data are in minutes):

Procedure	*Mean*	*N*	*St. Dev.*	*Min.*	*Max.*
CT scans	48	88	18.9	19	126
Specials	90	52	38.6	30	217
Ultrasound	16	48	7.1	4	30

15. Suburban's Nuclear Medicine Department was also a distinct organization that remained independent of the Radiology department over the course of the study.

16. When assigned to ultrasound and nuclear medicine, Urban's radiologists continued to dictate readings on films produced in the Main Department, but otherwise they rarely assisted with the Main Department's work.

17. After ultrasound and CT duty were combined at Suburban, sonographers began to complain that radiologists were now more difficult to summon in a timely fashion. Temporal data collected from Suburban's Ultrasound Department, however, suggested that there was no statistically significant increase in radiologists' response time after the scanner came on line. Rather,

it seemed that sonographers perceived that the radiologists were unduly interested in the scanner and as a result sonographers became more impatient with any delay whatsoever.

18. Few radiologists told sonographers that they held the sonographer's interpretive competence in high regard. To have made a point of the sonographer's interpretive competence would have been to publicly acknowledge that sonographers could interpret, a role traditionally denied to technologists. Among themselves, however, radiologists spoke of sonographers' abilities and several radiologists at both departments told me on numerous occasions that they thought their most competent sonographer was better versed in the modality than many radiologists.

REFERENCES

Barley, Stephen R. 1984. "The Professional, the Semi-Professional, and the Machine: The Social Ramifications of Computer Based Imaging in Radiology." Ph.D. diss. Massachusetts Institute of Technology.

———. 1986a. "Changing Roles in Radiology." *Administrative Radiology* 5.

———. 1986b. "Technology as an Occasion for Structuring Observations on CT Scanners and the Social Role of Radiology Departments." *Administrative Science Quarterly* 31.

Blau, Peter M., Cecilia McHugh Falbe, William McKinley, and Phelps K. Tracey. 1976. "Technology and Organization in Manufacturing." *Administrative Science Quarterly* 21.

Blauner, Robert. 1964. *Alienation and Freedom.* Chicago: University of Chicago Press.

Braverman, Harry. 1974. *Labor and Monopoly Capital.* New York: Monthly Review Press.

Child, John. 1972. "Organizational Structure, Environment, and Performance: The Role of Strategic Choice." *Sociology* 6.

Chinoy, Ely. 1955. *Automobile Workers and the American Dream.* New York: Doubleday.

Comstock, Donald E., and W. Richard Scott. 1977. "Technology and the Structure of Subunits: Distinguishing Individual and Workgroup Effects." *Administrative Science Quarterly.*

Cottrell, W. F. 1939. "Of Time and the Railroader." *American Sociological Review* 4.

Dewing, S. B. 1962. *Modern Radiology in Historical Perspective.* Springfield, Ill.: Charles C. Thomas.

Faunce, William A. 1958. "Automation in the Automobile Industry: Some Consequences for In-Plant Social Structure." *American Sociological Review* 23.

———. 1965. "Automation and the Division of Labor." *Social Problems* 13.

Hackman, J. R., and E. E. Lawler. 1971. "Employee Reactions to Job Characteristics." *Journal of Applied Psychology Monograph* 55.

Hackman, J. R., and G. G. Oldham. 1981. *Work Redesign.* Reading, Mass.: Addison-Wesley.

Hage, Jerald, and Michael Aiken. 1969. "Routine Technology, Social Structure, and Organizational Goals." *Administrative Science Quarterly* 14.

Hall, Edward T. 1959. *The Silent Language.* New York: Doubleday.

———. 1969. *The Hidden Dimension.* New York: Anchor Press.

Herbst, P. G. 1962. *Autonomous Group Functioning.* London: Tavistock.

Hickson, David J., D. S. Pugh, and Diana C. Pheysey. 1969. "Operations Technology and Organization Structure: An Empirical Reappraisal." *Administrative Science Quarterly* 14.

Larkin, Gerald V. 1978. "Medical Dominance and Control: Radiographers in the Division of Labor." *Sociological Review.*

———. 1983. *Occupational Monopoly and Modern Medicine.* London: Tavistock.

Lawrence, Paul R., and Jay W. Lorsch. 1967. *Organization and Environment.* Cambridge, Mass.: Harvard University Press.

Mann, Floyd C., and Richard L. Hoffman. 1960. *Automation and the Worker.* New York: Henry Holt.

Miller, Eric J. 1959. "Technology, Territory, and Time: The Internal Differentiation of Complex Production Systems." *Human Relations* 12.

Mumford, Enid, and Mary Weir. 1979. *Computer Systems in*

Work Design—the ETHICS Method. New York: Halsted Press.

Noble, David F. 1979. "Social Choice in Machine Design: The Case of Automatically Controlled Machine Tools." In A. Zimalist, ed., *Case Studies in the Labor Process.* New York: Monthly Review Press.

———. 1984. *Forces of Production: A Social History of Industrial Automation.* New York: Alfred A. Knopf.

Perrow, Charles. 1967. "A Framework for the Comparative Analysis of Organizations," *American Sociological Review.* 32.

Rice, A. K. 1958. *Productivity and Social Organization: The Ahmedabad Experiment.* London: Tavistock.

Roth, Julius A. 1963. *Timetables: Structuring the Passage of Time in Hospital Treatment and Other Careers.* Indianapolis: Bobbs-Merrill.

Schein, Edgar H. 1981. *Career Dynamics.* Reading, Mass.: Addison-Wesley.

Shaiken, Harley. 1979. "Numerical Control of Work: Workers and Automation in the Computer Age." *Radical America* 13.

Trist, E. L., and K. W. Bamforth. 1951. "Some Social-Psychological Consequences of the Longwall Method of Coal Getting." *Human Relations* 4.

Trist, E. L., and H. Murray. 1958. "Work Organization at the Coal Face." Tavistock Institute, London, doc. 506.

Van Maanen, John. 1977. "Experiencing Organization: Notes on the Meaning of Careers and Socialization." In Van Maanen, ed., *Organizational Careers: Some New Perspectives.* New York: John Wiley.

Van Maanen, John, Edgar H. Schein, and Lotte Bailyn. 1977. "The Shape of Things to Come." In L. W. Porter et al., eds., *Perspectives on Behavior in Organizations.* New York: McGraw-Hill.

Walker, Charles R., and Robert H. Guest. 1952. *The Man on the Assembly Line.* Cambridge, Mass.: Harvard University Press.

Wiener, Robert. 1950. *The Human Use of Human Beings.* Boston: Houghton Mifflin.

Woodward, Joan. 1958. *Management and Technology.* London: HMSO.

Zerubavel, Eviatar. 1979. *Patterns of Time in Hospital Life*. Chicago: University of Chicago Press.

———. 1981. *Hidden Rhythms: Schedules and Calendars in Social Life*. Chicago: University of Chicago Press.

CHAPTER 5

Janus Organizations: Scientists and Managers in Genetic Engineering Firms

Frank A. Dubinskas

"GENETIC ENGINEERING" companies are on the cutting edge of a new amalgam of revolutionary biology and entrepreneurial adventure in high technology. This amalgam was forged under intense pressures of professional hubris for molecular biology and high-risk, new-venture finance in the early 1980's. And it was forged from two hitherto largely separate social universes: the world of academic molecular biology and the world of high-tech entrepreneurial management. These were new organizations with two public visages—one of scientists and one of entrepreneurs—and, like the Janus of Roman mythology, these faces often spoke in opposite directions. As the two professional cultures were melded into a single organization—a start-up biotechnology firm—the new entity often experienced serious problems and conflict over their integration. Biologists and executives clashed, and still often argue over what goals research should pursue and how the choice should be made, or whether directions should change and projects be dropped.

The integration of managerial, marketing, and financial objectives with the scientific and technical productivity of the organization is a difficult challenge. This challenge is infused with a sense of urgency over the need to deliver real products from these costly enterprises, rather than just research and promises, before their investors and creditors pull out the lavish rugs of financial support. In this highly energized environment, cooperative planning toward joint, realistic, and realizable goals is hindered by a cultural gap between the two major groups of actors.

By "culture," here, I mean each group's shared and coherent (if loosely structured) ways of interpreting their worlds, and their patterned processes of communication and interaction. In this model, "pattern" and "process" are intimately linked. Whatever coherence or pattern is discerned is reciprocally linked to the social action through which it is manifest. In forging these new industrial organizations, one realm of cultural differences that is especially salient to company scientists and managers is time. The temporal aspects of project planning and work progress are a continuing source of disagreement and misunderstanding, according to both executives and biologists.[1] Each group has a different sense (or, better, senses) of temporality—culturally constructed understandings of what time is in different contexts of their working lives. These understandings inform and are shaped by different ways of patterning their work activities across a wide range of contexts. The differences are based in the fundamentally divergent kinds of technical expertise controlled and exercised by each group. The relevant contexts of difference examined here include developmental models of life-cycles and professional careers, work projects as diverse as creating a marketing plan or cloning a cell line,[2] the temporal order (or disorder) of daily work routines, or the image of time as a commodity to buy and consume.

In this study of cultural contrast and conflict in genetic engineering firms, two realms of time are particularly important. I call them "developmental time" and "planning time." By "planning time," I mean the temporal patterning of activities—their pacing, scheduling, and organization—in the future. Company planning activities are a crucial arena of interaction and negotiation between scientists and executives. The significant contrast between the groups here is "short term" or "near term" versus "long range" or "long term." The temporal construction of plans and the work that planning envisions and promises can be the sources of deep-rooted misunderstanding and conflict. Some of these roots lie in the contrasting images of selfhood that each group holds. That contrast turns around differing senses of what constitutes an appropriate life-course and professional career, and the images are cast as continuous movement (for scientists) versus maturation (for managers). "Developmental time" is my general gloss for these images of the self as a growing or maturing person, following a culturally *qua* professionally appropriate path—changing through the past into the present and anticipated in the future. This maturation or development evokes both "natural" and "social" processes simultaneously in the natives' rhetorics. In broader perspective, "planning" and "developmental" times are symbolic nexes around which important images coalesce of how the company world works and who populates it. As key symbols in the context of company planning and negotiation, these images of time are also *used* as rhetorical tropes in the process of argument over what and how things are to be done. That is, scientists and executives not only understand, inhabit, and enact their cultural patterns but also wield these temporal images as tools or weapons in an arena of contentious interaction. "Time" and "timing" are bones of contention as each group tries to argue for (and in) its own way in making company decisions. In sum, the primary anthropological project of this chapter is to examine how images of the self and patterns of work processes are articulated through the central symbolization of time, and how these generate or contribute to the reported conflict.

A further interpretive aim of my work is to account for

some of these differences between the executives and biologists by examining their contrasting histories of socialization. The two groups come to their companies with long histories of separate experience; they are imbued with their respective professional cultures for much of their lives. Those experiential histories include training and education through formal socialization processes that help shape their characteristic styles of understanding and acting in the world. In this research[3] I paid special attention to the development of cultural patterns in styles of pedagogy and interaction through formal education. I argue that each education cultivates a particular style of problem solving, that this problem-solving style is consonant with each group's own self-image and with its image of the fundamental nature of professional work, and that this cultural pattern is manifest in its rhetoric of argument in company planning.

FOUNDATIONS OF AN INDUSTRY: GENETIC ENGINEERING AND HIGH-TECH ENTREPRENEURSHIP

What are "biotechnology" or "genetic engineering" companies, and how did they come to be? I focus on an unusual young cohort of start-up entrepreneurial companies, most of them located near Boston, San Francisco, and Washington, D.C. The first company, Cetus Corp., was founded in 1971; but the new entrepreneurial wave really began to form and spread after 1976. By late 1984, there were about 200 such firms, the majority of which had been founded since 1980 ("GEN Guide to Biotechnology Companies," 1983; OTA Report, 1984).[4] While large pharmaceutical, chemical, and petrochemical firms are also now players in the biotechnology game, my research is focused on small companies, as new interfaces between the worlds of high-risk finance and academic molecular biology.

Pervasive, deep-rooted patterns of difference are often

most evident at the initial stages of the high-tech start-up. As companies mature, they develop varying strategies for dealing with conflict, and as the industry matures, these strategies become more widely discussed and distributed. Examination of this process is a focus of my ongoing research, but this essay concentrates on the earlier years—and the more obvious contrasts evident in new firms in the throes of birth into the business world.

There are at least four enabling circumstances or forces whose historical confluence engendered the meteoritic burst of small biotechnology ventures onto the business scene in the late 1970's and early 1980's. These are a transformation of the venture capital industry, a revolution in biological techniques and understanding, a re-emergence of the worship of entrepreneurs as heroes, and finally the explosion of government-funded research on a class of obscure biological substances—interferons.

Adventure Capital

New business collaborations need money to get started, and, in the late 1970's, the financial world saw a resurgence or transformation of its high-risk venture capital sector. For a decade before then, high-risk venture capital was a tiny segment of the investment banking industry, mostly dedicated to handling a small portion of the immense personal wealth of a few rich families. Another segment of high-risk venture banking had been heavily involved earlier in leveraging the merger boom of the 1950's to the early 1970's, but these financiers were *not* accustomed to setting up companies *de novo,* and certainly not in emerging high-technology fields.

By 1978, a decade of drought in venture funds was ending, as two major changes in tax laws and investment regulations occurred. The first was reduction of effective capital gains taxes from nearly 50 percent down to 27 percent, then to 20 percent under the Carter administration. The second was the easing of regulations governing the investment of monies held in trust. This loosening now permitted a small proportion of the great industrial pension funds to be allocated to high-risk ventures. Even a small part of those vast resources was a lot of money, and, by 1980, there was a flood of cash (including some from overseas) looking for a place to invest. A partly new and perhaps less experienced cohort of bankers—today's venture capitalists—emerged to manage and invest this fortune. They were armed with the best modern business-school tools—high risk for high return was their specialty—and they cast about for adventurous new ways to invest. And they found the biologists, who promised to transform pharmaceutical and chemical production with their new genetic engineering technologies.

The "New Biology"

Genetic engineering, as a practice, was enabled by revolutionary new work in molecular biology in the early 1970's. Techniques were developed in university research labs that allowed scientists to alter the hereditary material—the "genome"—of living organisms. DNA, or deoxyribonucleic acid, is the primary encoding material for genetic information in nearly all species.[5] In "genetic engineering," loosely described, one clips a segment of the DNA—usually a single gene—from one kind of organism and splices it into a completely different host organism. This host's newly augmented DNA is called "recombinant DNA," or "rDNA." The new gene added to the host's DNA now expresses a trait formerly unknown to the host. Furthermore, this trait continues to be expressed in the host's progeny, as it divides and reproduces.[6]

The commercial production of human insulin in a bacterial host, *E. coli* (common gut bacteria), is an example of the practical consequences of this "gene-splicing" technology. The two genes that encode for insulin production in humans have been identified and isolated. Rather than excising the relevant DNA from its human chromosomal location, in this case "artificial" genes (complementary DNA strands) were synthesized chemically. This DNA was then spliced into a plasmid "vector"—a means of getting DNA from one cell to another—and introduced into a host bacterium. The vector then inserts and merges its DNA into the DNA of the host. This bacterium now has the genetic capability to produce insulin. Bacteria reproduce themselves far more rapidly than humans (and with no incest taboos to worry them), and their newly inserted DNA instructs them to crank out insulin as a byproduct of their own metabolism. Developed by Genentech in San Francisco, the commercial product Humulin is now produced and marketed by Eli Lilly, a large pharmaceutical company.[7]

The exciting promise of this new technique is that microbes can be turned into tiny, specially designed factories for the production of anything from rare biologicals and pharmaceuticals to industrial organic chemicals. This is not to mention the prospects for agriculture in altering the genetic character of livestock, plants, and their symbiotic neighbors, or in a host of other fantastic potential projects that the biologists and entrepreneurs can invoke.[8]

Entrepreneurial Heroes and Heroic Biologicals

The basic biological work on gene splicing and cell cloning was widely publicized and disseminated by the 1975 Asilomar conference on recombinant DNA and its potential hazards.[9] That peak of public controversy over biological hazards and the ongoing one over environmental alterations have contributed to the urgency and energy around the industry. Shortly after Asilomar, the first industrial research collaborations were established. Entrepreneurs and venture capitalists were seduced by the commercial promise that the scientists implied; and some of the scientists also sought *useful* products, partly to redeem the public distrust over rDNA development.[10] The new commercial collaborations brought top-flight university research scientists from molecular biology (mostly) together with entrepreneurs and venture capitalists, who founded the new genetic engineering companies.

This kind of commercial collaboration was relatively novel for academic biologists. While many of them had done some industrial consulting before, suddenly they were being buttonholed by investors who saw their expertise as a key to the next big high-technology breakthrough. The models of the microelectronics and personal computer industries were often compared with the economic potential of the new biotechnology, and high-risk investors wanted to get in on the ground floor. At the same time, a change was taking place in the social climate of the United States. Entrepreneurship, particularly in high technology, was touted as the new savior of American prosperity. Entrepreneurs were the Davids of innovation, fighting the bureaucratic Goliaths of failing industry. Steven Jobs and Steve Wozniak with their Apple™ computer are only the best-known of this phenomenon. And biologists were partly warranted by this broad cultural change to establish business partnerships and equity positions in the new start-up companies. The profession made a major shift from general skepticism or censure about Herb Boyer's first step into Genentech in 1976, to widespread approval of and participation in firms by top biology faculty members by the early 1980's.

A final important actor in this drama is interferon. Sandra Panem, in *The Interferon Crusade* (1984), has dismissed in detail the politics of funding and hype surrounding this once relatively obscure family of human immunoregulators. Massive research interest in the substance(s) was peaking just as the new biotechnology ventures were setting up, and there were "indications" that interferon(s) might be a cure for anything from cancer to the common cold. Interferon was seized upon by the new industry as a heroic "big hit" product, which could prove the worth of the new industrial enterprises. The top academic molecular biologists probably had a hand in this strategy, since they were from professional ranks and institutional environments where going for the big (Nobel) win or discovery or theory was part of their professional ethos. But more about these scientists in a moment.

The process of building these entrepreneurial collaborations has been a very rocky and complicated one, both as science and as business. The steps from the scientific lab bench to a packaged product on the shelf are not trivial, and neither are they simply implied in the molecular genetics and microbiology of the research environment. The production problems in scale-up from the petri dish to the thousand-liter fermentor are vastly more difficult than the difference in scale.[11] Besides this, a whole universe of new patent protection, safety, marketing, and distribution problems has barely begun to be addressed in the companies. Many potential products, especially human pharmaceuticals with their long regulatory testing tracks, will require nearly a decade from the first establishment and purification of a substance in the lab to the day when a saleable product is out the door. That is a long time to wait for profits—far longer than a venture capitalist's common expectation of three to five years before cashing out of a firm. There is continual pressure from the managerial and entrepreneurial cadres of the companies, as well as from investors, to force the scientists to focus and accelerate their work. This often translates into pressure to find short-term "doable" projects[12] that can generate quick revenues. After three or four years, urgency increases as executives seek further rounds of funding from investors, and this urgency is conveyed down through the organization to scientists. Entrepreneurs must present compelling accounts of their successes (or potential ones) to outside investors, and the scientists' work must somehow form the basis of these accounts. I imply in this that there are two "sides" to the companies in these early stages of their histories, and that is a model that I will now take up in more detail.

Scientists and Managers

A general characterization or global map of these start-up companies suggests an initial internal division into two fairly distinct culture-worlds: the "scientists" and the "managers." When I say this, though, I mean a shorthand for these particular kinds of biologists and entrepreneurial executives. This whole book points to the diversity of cultural variants in sciences and technologies. At the earliest points in their histories, the common divisions of industrial firms and business organizations into functional groups (like production, operations, personnel, marketing, or sales) barely exist. Most new companies have 15 to 40 employees; and only a few of them exercise managerial, financial, accounting, human resource management, staff support, and proto-marketing functions. Even in the "large," older firms like Biogen, Cetus, and Genentech with 400–700 or more employees each, non-research staff are still a small minority compared to the bench scientists and laboratory support personnel. A 20 percent to 80 percent split would be a good rough characterization for many small firms. The majority of employees are laboratory scientists and their helpers; PhDs and postdocs in "wet biology" (experimental laboratory biology, as opposed to, say, "field biology"), laboratory technicians, dishwashers and animal care staff, and a few technical support specialists like computer software engineers, electronics technicians, glassblowers, or machinists.

Few top university scientists—the ones whom the venture capitalists recruited as "founders"—actually became operating heads of their firms. Dr. Walter Gilbert, the Nobel prize-winning Harvard biologist and former Biogen CEO, was one of the last exceptions to this rule. Even he eventually (in December 1984) quit his executive position at Biogen (Bulkeley, 1984) in favor of a more "traditionally" experienced chief officer. In the firms studied, such distinguished biologists usually sat on the boards of directors of their companies, or on their "scientific advisory boards," while they also continued to teach and do research in their universities. Many start-up firms were initially modeled on a university research laboratory, and the bench scientists were recruited from among university postdocs with the promise that some proportion of their time, even a "fixed" proportion, would be reserved for "pure" research in their own field of intellectual interest.[13] Biologists were also commonly offered two to four times their university pay (at whatever level), a splendidly equipped lab, and an equity position in the firm (partnerships, stock, or stock options).

By contrast, many of the early management teams in start-up biotechnology firms were headed by executives-*cum*-entrepreneurs from the world of finance, and trained in top American business schools like Stanford and Harvard. They often had little or no current background in the biology upon which their firms were based, but came rather from the ranks of professional entrepreneurial managers, managers who specialize in high-technology startups and their financing. Their training and experience was often in finance, especially investment banking and venture capital, or occasionally from marketing, rather than, for instance, R & D management, manufacturing, or operations management departments in a related industry. This makes the biotechnology industry unusual among either established technology-based industries (like aerospace) or newer high-technology ones (like microelectronics). In those industries, managerial cadres are most often promoted from the ranks of engineers and scientists. By contrast, start-up biotechnology firms seem to have a relatively high proportion of non-scientists, business-school-trained managers, whose primary expertise is in dealing with the outside financial world of investors rather than the internal management of research and development operations. Besides this, since the technical and scientific staff of the new companies were recruited directly from universities, there has been no internal, industrially experienced cadre to call upon for promotion into managerial ranks.

I call all of a nascent firm's managerial/entrepreneurial group "managers," even though a few company founders are entrepreneurial scientists. Whatever their origins, in the company context, this group is motivated primarily by managerial and business concerns of the firm, rather than by the professional concerns that motivate the practicing bench scientists. As one biology professor and new company founder opened his address to a seminar on biotechnology opportunities: "There are three major points to keep in mind, as you begin to formulate your [business] plans: The first is money. The second one is money. . . . And the third is money, too! [laughter]" These new managers, like their MBA-trained counterparts, largely identify themselves with a business community and are seen as such by their own scientific staffs.

It is important to emphasize, here, that the cultural patterns I describe are not innate properties of the people who exhibit them, nor do individuals only/always "belong to" a single cultural variant. The professional-culture patterns I describe are more like repertoires of coherent possibilities for and permutations of action than like "rules" of conduct or value-models. Each cultural variant permits choice and negotiation among an open-ended set of alternatives, which still appears as a loosely consistent, patterned system. Furthermore, individuals are not "glued" to any one particular cultural variant; and some people can obviously move between different systems with facility. Scientist-entrepreneurs are a telling example, since they *are* appropriate persons in each of two contrasting culture-worlds, depending upon the specific setting in which they act. As faculty members in their universities, they *are* scientists; and as company founders, they *are* entrepreneurs. Their "selves" are constituted in the different arenas of interaction in which they participate.[14]

One significant and growing minority of biotech start-up executives is a partial exception to this bipolar model of staffing contrasts. These are scientist-managers who are recruited into small firms from large pharmaceutical and petrochemical companies. Usually they are PhD scientists who made a move from the bench into the management of research or new product development divisions at industrial giants earlier in their careers. With successful track records of managing industrial science, some of these big-company R & D heads have been recruited to lead biotech start-ups. Their industrial experience is sought in order to introduce "practical" management and control to the "blue sky" efforts of molecular biologists fresh from the academy. Their specific experience, though, is as heads of one small function in a huge organization; and their relatively constant budget, at perhaps 3–5 percent of the parent's annual expenditures, still exceeds the assets of most new biotech ventures in both size and stability. It is not clear whether the managerial skills and expertise of these executives will suit the high-pressure, high-risk, and highly "academic" ethos of new biotechnologies companies. While their PhD backgrounds may help them to interpret contemporary biology, they have usually left day-to-day experience with techniques and developments in the field far behind. Like the new entrepreneur-scientists fresh from academe, these professional "science managers" from industry are stepping into a new and rapidly changing environment. But like their counterparts from the world of finance, they are clearly in the business camp of nascent firms.[15]

To return to the scientific personnel, bench scientists within start-up biotechnology firms usually are recruited fresh from the laboratories of top university academic departments, rather than from the ranks of practicing industrial scientists. This is partly because firms seek to exploit the cutting edge of scientific work in the new biology, and this work is being done primarily in the government-funded laboratories of prestigious research universities.[16] Second, these firms are usually set up with the close collaboration of a relatively high-ranking university scientist, who looks within her or his professional networks of associates to fill positions. Persons of known qualities are preferred to strangers, if the people are of the same experience and (presumed) caliber. Scientists recruit from among their own or their close colleagues' PhD students and postdoctoral research fellows (postdocs), and even into the technician level, for people with the appropriate expertise to staff companies. Some small firms can trace nearly their whole staffs from the "professional kinship" networks of the founding scientists:

> At one young California company, a PhD biologist—a section head—counted through the dozen and a half people at his site. There was a lone finance VP and a secretary, and the rest were scientific staff. All but two count their professional genealogy through one of the two faculty founders—professors at nearby universities. These founders sought recommendations from former teachers, students, and present close colleagues; or they hired their own students and postdocs. Except for two "distant cousins" with special

> expertise scarce in the closer group, everyone had some close link to two university labs.

Scientist-founders usually say that their ventures will depend for success on the quality of their scientific staff, and that they know best how to evaluate people that they have some personal knowledge about. The oral exchange of reputations through personal networks is an important part of building scientific communities (see also Traweek, 1982a), and that same networking stvle is imported into early company personnel decisions.

This network-based recruitment seldom reaches into the ranks of established industrial scientists during early company histories. The two groups probably remain separate even when biotechnology companies grow to the size (above about 50 employees) when they begin to bring in more staff through public advertising (e.g., in journals like *Science*, *Nature*, and *Bio/Technology*). University faculty in molecular biology and related fields have a strong tendency to see their own work as far more technically advanced as well as intellectually more potent than anything done in industrial R & D labs in large companies. Although many faculty-founders do (or did) consult to big petrochemical and pharmaceutical companies, they always come in as the high-status outsiders to ponder specific technical questions, and they seldom develop intimate working ties with bench scientists there. They tend to deprecate the "quality" of the science done anywhere but in university labs—even including their own commercial ventures, on occasion. Small firms recruit their scientists straight out of universities, and this personnel brings with it many of the work patterns, styles of interaction, values, and expectations to which they were accustomed in academe. As one young postdoc said: "[The company]'s just like a university lab. . . . We go up to campus for seminars." The unspoken but obvious statement was her strong identification with the campus ethos and laboratory life-style that they had left just months before.

The managers and scientists of young companies identify themselves as being in—and "from"—different "worlds," and a number of visual symbols point to these differences. Separate identities are distinctly marked in the company environment through contrasting styles of dress, which are immediately obvious to see. Scientists tend to preserve their customary "laboratory casual": blue jeans or slacks and sport shirts for men and women, with occasional skirts for women. Functional group heads (e.g., for a "protein chemistry" lab group) may keep a jacket and tie handy for meetings with the higher managerial staff, but only at the research director's level does that dress seem common. By contrast, the business and managerial group tends to suits and ties (grey pinstripe on the East Coast) for the men and business suits for the women. Support staff in the managerial area dress more formally too: men with jackets and ties, and women in dresses and heels. Crossing the geographical *qua* social lines within a firm makes one visually obvious at a glance as coming from the "other" side.[17] Employees are acutely aware of these differences, and they are the subject of much deprecatory humor by each group about the other. A board member *qua* academic scientist was almost embarrassed in explaining away the business suit he was wearing in his university lab one day: "I have to chair a board meeting today, so I can't go in [to the company] looking like I usually do around here."

Another visual marker of group differences lies in the kinds of pictures and images that grace the walls of labs versus those of executive office areas. A mahogany row or managers' corner of the building is often decorated with designer-coordinated "modern art." These are abstract patterns on large canvases, usually in a fairly uniform style throughout the managerial area, which may be bought or rented (and thus changed on a regular basis). An alternative is the occasional specially commissioned work or mural, like the ones in Cetus Corp's old Berkeley executive headquarters that depicted company themes. Rented plants also abound in these areas. Working laboratory areas, by contrast, have motley collections of photos, posters, humorous cartoons, and notices on a variety of bulletin boards, door, and walls. This apparent disorder has some consistencies, like the prevalence of "outdoorsey" and nature posters, photos of lab members and their social activities, humor that pokes fun at the scientists and their work (visual and written), and occasional representations of data. Executive offices, foyers, and meeting rooms seldom have any personalized decoration (like informal staff photos), nor do they sport the collage of shreds and patches from the everyday lives and work that labs do. Old spreadsheets and business plans do not grace their walls. But behind these visual boundary markers lie years of learning how to be an appropriate member of each group—how to do one's particular professional work.

THE WORK OF SCIENTISTS AND MANAGERS

The fundamental differences between these executives and scientists lie in the distinctive knowledge bases and practical expertise of each group in its own field. The meat of these differences lies in the character and conduct of everyday work—its technical detail. Rather than writing an exposition of that immense complexity, however, I have focused on one *joint* activity—project planning—and used it as a point of entry into these worlds of difference. The technical "content" of work is also reflected in its style of accomplishment, and the styles of accomplishing "managerial" versus "scientific" work are different. By "style," here, I mean the consistently observable patterns and coherence in everyday, practical activities, including verbal interactions, of each group. Parameters of difference include arenas and contexts of interaction, visual appearance, styles of cooperation, relations to "contestible" information (Traweek, 1982b), relations to hierarchy and/or authority, rhetorical tactics for establishing or attacking authoritativeness, and the key metaphors in which issues are cast. One of these key metaphors—in fact, a key symbolic nexus—is "time"; and it figures in the crucial areas of contention over planning and decision making that motivate this chapter. Temporal pattern also informs the distinctive images of self in each group, as well as their broadly understood images of "science" or "management" as professional activities.

Managers write, phone, manipulate electronic data images, plan, and try to raise money; they meet and talk with others to gather information, form opinions, share gossip, and make decisions. They mobilize and direct managerial staff and scientists alike. This work is embedded in complex social interactions involving argument, exposition, and emotion in the contexts of their "technically" economic, fund-raising, managerial, and planning work. The classic observations by Henry Mintzberg (1973) also show that their activities are temporally fragmented in the extreme, with executives rarely spending more than 15–20 minutes at one activity, unless it is a scheduled meeting.

Scientists read, confer, mix chemicals and biological substances, prepare gels, operate experimental and recording devices, examine the electronic and paper traces of machine outputs and computer programs, write on paper and on sample containers, and compose reports, articles, and grant proposals. Their professional work is likewise a densely textured interactive process. There is a growing body of ethnographic research conducted in and on scientific laboratories, often focusing on the details of practical bench work. Latour and Woolgar's (1979) study of a neuropharmacology lab at the Salk Institute in LaJolla, California, is a classic of the genre, followed by Knorr-Cetina's

(1981) study of a biology laboratory at UC-Berkeley.[18] The particular activities characteristic of each group—managers and scientists—are far too extensive to treat in detail here. Furthermore, representing them briefly as a list of different "tasks" seriously damages the sense in which they are intimately interconnected in practice, a characteristic of professional work that is made evident in two genres of research: laboratory studies and R&D management studies.

Laboratory Studies

Detailed studies of laboratory and management practice can contribute immeasurably to our ability to interpret their separate (or overlapping) spheres of activity by examining the micro-processes through which, for instance, contentions are transformed into "facts." While this work is well underway for various sciences, it is just beginning to be addressed for technologies (see, e.g., Pinch and Bijker's 1984 programmatic article), and it is so far virtually absent from management studies.[19]

The early and seminal study by Bruno Latour and Steve Woolgar (1979) at the Salk Institute is of special relevance for studying genetic engineering firms. The everyday scientific work that they examined at Salk is closely akin to many laboratory practices in molecular biology. Among other tasks, the authors deconstruct the historical process of establishing a scientific "fact": the existence (or discovery) of something called TRF(H) (tryptophan releasing factor/hormone). They describe the micro-processes in laboratory conversations and everyday work activities through which this "fact" is constructed. They show convincingly that the activity that we think of as "scientific lab work" consists of mutually intertwined activities where "beliefs are changed, statements are enhanced or discredited, and reputations and alliances between researchers are modified" (pp. 157–58). Verbal exchanges in the lab are heterogeneous in their mixtures of what we might call economic, social, political, temporal planning, professional, and "technical" references as *simultaneous* components of actions and interchanges. For example, "in the course of one short discussion references are made to subject matter, to personalities, to claims made at a conference, to techniques used in another laboratory, and to competitors' past claims" (p. 165). This activity is what we have previously rather prosaically glossed as "technical considerations" or "technical science." Embedded in it are a wide spectrum of "social" considerations, and they are an intimate and inseparable part of the way that scientific work is built. It suggests that a similar attention to the detail of managers' work would uncover the processes through which they also build the facticity of their "real (*qua* economic) world," too.

R & D Management Studies

Anthropologists are certainly not alone in trying to understand scientific and technical management decisions. Decision analysts, cognitive psychologists, systems analysts, sociologists, economists, social psychologists, strategic planners, and other management professionals have all been working in the field for a number of years. Many of them are trying to develop or systematize work and management processes into "rational" models, with some sort of logical, sequential order that can be applied to a set of similar cases. Other work, however, flies in the face of attempts to totally "rationalize" design choices and scientific activity. (Chapter 3, above, is an example of a new synthesis in this direction of ethnographic and engineering management approaches.)

Major new work began to appear in the mid-1960's, much of it stimulated by Tom Allen's now classic studies on communication in R & D environments. Working out of MIT's Sloan School of Management, Allen studied a variety of firms and government agencies, including one particularly instructive example for us. NASA (National Aeronautics and Space Administration) engineers were trying to evaluate and choose among three possible designs for a large antenna subsystem (Allen, 1966, 1977). Allen monitored engineers at two labs working on the same project. One of Allen's diagrams (1966, p. 75, fig. 2) for one of these labs is partially represented below in Figure 5-1. The three small symbols represent the three different technical approaches to a solution; the vertical axis plots the relative probabilities (0.0–1.0) of choosing each of the three approaches; and the horizontal axis plots the time (in weeks) at which the group leader ranked the probabilities. Each week, Allen asked group leaders to rate the probability that each of the three alternative technical approaches would be their eventual design choice. These relative probabilities were mapped over a 36-week period, until the design choice was settled in each group.

The "maps" or diagrams of projected choices shows extreme fluctuations in the relative "favor" of different designs from week to week. Designs soared and swooped up and down in probability, until a final choice emerged. What is notable for us in this chart of choices is its apparent "messiness"—its lack of predictable order in the twists and turns evaluating each choice. Allen showed, through retrospective interviewing of group leaders, how this seemingly erratic indecision actually reflected a continuing process of balancing technical and economic considerations in concert over the three design styles, until a consensus choice was reached. This intimate intermixing of "technical," "social" (*qua* economic), and "historical" considerations is precisely the same process described in more conversational detail by, for example, Latour and Woolgar (1979) in the neuropharmacology lab or Woolgar (n.d.) among solid-state physicists.

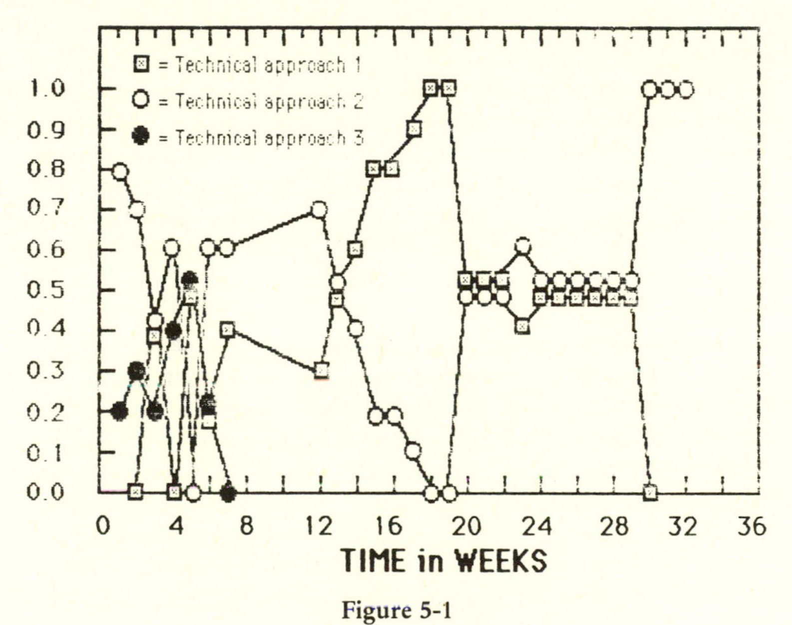

Figure 5-1
Laboratory "A"
Adapted with permission from Allen (1966), p. 75

When the decision was made, however—30-plus weeks down the road—Allen asked the engineers how they had come to make their choice. They universally responded with a story or "account" of the process that cast it as a unilinear sequence of progressive, logical steps leading only and inexorably to the actual choice made. This account "erased" all of the uncertainties of the process and turned it retrospectively into a logical, "scientific" story of why their choice was the best—and, in fact, the only possible choice. One lesson is that retrospectively constructed accounts commonly obscure or ignore the social processes through which choices are negotiated. They selectively represent that work, be it of engineers or scientists at the bench or of managers at the desk. Scientific papers and company annual reports are two excellent examples of where this kind of account is inscribed and then used in other social contexts, like reputation building and fund raising.[20]

In genetic engineering companies, the social processes of managerial and scientific work go on largely in isolation from one another. We can model each as a relatively autonomous practice, each with its own dense, complex interwoven sets of negotiations and considerations working toward outcomes. Once results are "decided" or "discovered," each group constructs—with its own cultural patterns—a retrospective rationalization of its work. These "logical" accounts justify (or "account for") their outcomes in the most authoritative voice of each cultural variant. These accounts or stories about work are extensively circulated within each group, and between them—often through the media of "linking" managers like R & D directors. In the discussions at planning meetings, these retrospective accounts are then used as resources in arguing for one or another choice about what is to be done. The "when," or the temporal structure, of plans is one of the flexible characteristics of many choices, and we turn now to a cultural interpretation of these "time" resources.

CULTURAL ANTHROPOLOGIES OF "TIME" AND PROCESS

Sharon Traweek's work on the experimental particle physics communities of Japan and the United States (1982a, 1982b, and this volume) is an example of how we can elucidate cultural patterns as global characterizations, while still attending to the details of physicists' lives and practice. Her decade-long work in the physics community has blocked out their social construction of "time" or "duration" in elegant complexity. "Time" has a vast array of significant permutations for physicists, and many relevant arenas or contexts of temporal organization intersect with those of scientists (and managers) in biotechnology firms.[21] The overlapping possibilities include "historical" time or constructions of the past of physics and molecular biology; careers, life-cycles, and development times; and cohort histories, as the development of a research group experiencing significant events together. There are also the cyclical times of years, weeks, and days, and of fiscal years and quarterly financial reports. There are the work-tempo alternations "uptime" and "downtime" on particle detectors and the generation times of experimental organisms and cell lines for the biologists. There are cycles of grant application deadlines and of academic calendars. Time is inscribed as a line on recorder paper for experiments (see also Woolgar, n.d.) and on a chart as project planning "milestones"; and it is "written into" business plans and research proposals, and treated as a commodity that can "run out" or be bought and consumed.[22]

"PLANNING TIME," THE LONG AND SHORT OF IT

One critical arena where the cultural construction of time is salient in biotechnology companies is in various kinds of planning activities. These involve the setting of goals and schedules for completing projects, especially schedules with internal markers of progress, called "milestones" in business and engineering parlance. It also includes planning on a larger scale, like the preparation of "business plans," documents that chart the future activities of the company across a wide front. While business plans are occasionally construed as "real" guides to company action, they are quite maleable in practice in these fast-changing firms. Short-term (quarterly and annual) plans are perhaps more likely to reflect programs toward which a company is actually working than are longer-term plans. Long-term business plans, in volatile biotechnology start-ups, are most often written on the occasion of seeking investors from the outside; and, in nascent firms, their flexible character is usually recognized by everyone involved.

In discussing "planning time," my object is to look at some more general characteristics of how managers and scientists order their projects, as "projections" of activity they can or will engage in the future. How "time" is construed in the environment of planning is important, since different images and understandings of "time" are the underpinnings of this scheduling of promises. In joint planning environments, the structure of a plan or schedule is negotiated within a group that includes both scientists and managers. Contrasts in what constitutes a culturally appropriate plan make the process of resolution and agreement more difficult. The basic contrast is between long- and short-term planning, a contrast that is always relative to the particular subject or activity being planned. This difference also maps onto a difference between "open-ended" and "closed" systems (a distinction that will be made clearer in the next section).

Managers see themselves as capable of both long- and short-range planning, but they often complain that they have "no time" to do long-range or strategic planning. They stress that immediate concerns like quarterly financial reports must take precedence over more "distant" ones like, for instance, five-year plans. They speak of this as an aspect of their native managerial "realism," especially in contrasting themselves to their scientists. Creditors and investors have a palpable presence in these plans, and imminent economic pressures foster a focus on the immediate present and proximate future. As one company board member said to me about contracts for research with big firms: "They [the company managers] *know* that milestones have to be met. Those businessmen negotiated the deal, and they *must* be able to explain to a partner why they haven't gotten to a goal." The *next* milestone, the *next* contract, and the *next* stockholders' report are the "real" near-term subjects that a manager must always address, and managers' self-images as "hard-nosed realists" emphasize their association with that immediacy.

Managers, on the other hand, see their compatriot scientists as having a very different world view about "time." They accuse scientists of not ever having an end in sight. Two managers had these comments: "You have to watch out for scientists out of control—every scientist thinks their work will 'make the company' " and "The biggest problem is a lack of focus. . . . There's this enormous technology engine which is roaring, and all these young, energetic people [scientists] . . . but you can't *afford* to do diverse experimentation." The range for scientists' plans, if they are even "plans" at all to the manager, is too long. It is too long, and it encompasses a seemingly infinite regression of possible branches and paths into an indistinct future. The scientists can't be easily pinned down. This open-ended approach to planning (or a struggle to keep projects open-ended) is viewed as unrealistic, in terms where "realism" has necessarily to do with short-term, and often economic, considerations.

Scientists, within this realm of planning "range," have quite a different way of characterizing themselves. They say that they can see (or "set their sights upon") far-distant and perhaps indistinct goals. The open-endedness of their temporal frames is construed as appropriate for their professional scientific work. They argue that science is drawn continually forward by the questions posed to it by nature (or occasionally, by science's own "internal logic") and that, in either case, there is no fixed end in view. Their work is *pro*spective in its practice and planning. Hence, management demands for milestones on research projects—definite stages of completed work toward a fixed, specific result—may be said to contradict the very essence of their work toward "discovery." The facts or answers are hidden out there in nature, so blocking out every stage of their discovery path in advance is futile. Scientists thus claim to have the long view, but deny that they can "plan" through it in the style that their managers require.

In conversation, however, scientists also present an artful solution to this dilemma on some occasions. While arguing that science, in general, fits the model above, they have two ways of finessing the planning problem. One is to argue that commercial science is a special case. They say that work in the firms is "more like engineering" or "technical" work than "pure" science. The distinction, to them, is like cranking out results from an established scientific paradigm—the puzzle-solving aspects of "normal science" that Thomas Kuhn describes (1962, pp. 35–42). They see this as "applied" science, which may be goal-directed in a way that "pure" science is not. For them, pure science builds theories, while applied science generates knowledge in new realms of investigation using mature theories.[23] It is particularly the scientists who consult to industry who seem to favor this argument, especially in the context of a discussion of their roles as outside advisors.

The other approach that scientists employ is to argue that the craft of really good science lies in choosing problems that are within the realm of solvability, not just in making brilliant breakthroughs. This image, as we will see, is not unlike the kind of balancing that PhD students and their advisors describe in trying to find a manageable thesis topic. In this case, the scientist, as a prospective seeker, acknowledges the infinitely unknown character of possible problems, and images scientific knowledge as being something like an envelope of relative certainty. Scientific work then consists of extending the boundaries of this envelope by launching new research from a platform or environment of past experience.[24]

Scientists conversely see managers as mired in short-range business concerns and limited by being "short-sighted." One important consulting scientist described his encounters with marketing managers in these terms:

> It's the marketing people . . . who don't understand how the *process* of science works. . . . Their sense of the world of [scientific] possibilities is too superficial. The scientists are looking five years ahead at what is being or *will be* developed, while marketing people are . . . rooted in the existing present state of perceived needs. They're just always stuck in the present.

Scientists see the goal of producing scientific knowledge as primary, and they tend to devalue economic goals in their world view. There is an aesthetics to this appraisal, too, where open-ended scientific work is more pleasing or "elegant" than the accomplishments of business ends motivated by short-run goals. Nobel prizes have more *panache* to them than veterinary dipstick diagnostics. When business executives complain that they have no time for long-range planning, scientists may accuse them of incapacity to do so, making that trait a generic aspect of managerial character rather than happenstance.

This should not strike us as an unusual critique. The current argument over corporate takeovers and the struggle between short-term profit taking by investors versus long-term capital investment is just the latest public splash over the issue. Many social and economic commentators have decried the "short-sightedness" of American business planning, particularly in comparison with Japanese industries. Executives, too, complain of this pressure—then often "rationalize" it. A biotech CEO once complained to me, when I asked how frequently he revised his long-range plans, "Long-range plans!?! I can barely get past the quarterly reports to do an annual [report]! . . . But the short run is what *counts.* If you don't make it in the short run, the long run doesn't matter."

In all these instances of self-description and the interpretation (or caricature) of the Other, the context of discussion should be kept in mind. This is the interpretation of a "problem," "contrast," or "difference" acknowledged by company personnel and seen by them as at least troublesome, if not critical. By setting up the terms of discussion in this way, a contextualized set of interpretations is forthcoming. This is not to say that the conflicts are not "real," just because the discourse through which they are created is contextually situated. Rather it is a reminder that all meanings are meanings-in-context. The understandings by each group of themselves and of the other are not immutable characteristics of the subjects so described, but rather occasioned interpretations by each group that are at least partially motivated by a need to establish authority and rectitude in the other's presence. The stakes are jobs and investments, reputations and careers, power and wealth, and satisfaction—quite "real" considerations for both groups, and also often quite different in their specific content. The context of this interpretive activity is a realm of choices in company direction(s) where each group believes that it should guide the company to its own goals, and fears that the other group may threaten to keep it from achieving those aims.

Developmental Time

Another kind of time that appears both as a part of these world views and as an arena for conflict is developmental time. By this, I mean images of the self (and occasionally the organization) as a growing, maturing organism. This image of maturation or development evokes both "natural" and "social" processes and metaphors simultaneously. This sense of development overlaps with the understanding of time in planning through images of the "complete persons" (for managers) versus "open-endedness" of, for instance, career plans for scientists.

Managers see themselves as rather unproblematically adult and fully formed human beings. Their careers may continue to "advance" throughout their lives, but their essential nature as "complete" human social beings is certainly accomplished by the stage of their first responsible executive position. Their advances involve getting "more" of what they already have—responsibility, authority, prestige, and financial reward—rather than a fundamental change of character. When they talk about "what is at stake" in their jobs, it is the growth of the firm (and hence their own responsibility and prestige)—and with it, the money. "You've got to keep your eye on the numbers. . . It may seem pedestrian to the scientists, but it's always a numbers game."

Managers often speak of scientists, though, as "immature," "adolescent" in their attitudes, or "not grown up." Staff at Biogen, for instance, referred to a group of postdocs whom their ex-president Walter Gilbert brought to the firm from his Harvard lab as "the hobby group." Managers connect their own maturity with their definition of the real world, one that is full of finite economic and "productive" activity. Productivity, for them, is usually measured in deliverable goods, rather than insubstantial and "unreal(ized)" ideas. As a company biologist said of them: "When they say *real*, they mean *money!*" The finiteness of produced objects is contrasted to the scientists' *in*finite capacity to expand knowledge, and the latter is often likened to children's play, with a deprecatory tone.

The biologists, on the other hand, characterize their own developmental processes as continuous throughout their lives. There are progressive stages in their education as undergraduates, graduate students, then postdocs. Getting a first faculty position at a research university is only the beginning of a trek through promotions, tenure, and increasing administrative responsibilites, both in the school and in their professional organizations. Finally come institute directorships, deanships, and "spokesperson for a field" status. Eventually one graduates to near-guru status, as an emeritus progenitor of a subfield or field.

This growth should never come to "completion." The growth is also located in the intellectual realm, the realm of productive "ideas," which are less valued as things-in-themselves by the managers. As one moves further along in status and political power, one is allowed more rein (especially in the latest stages of a successful career) to "philosophize" on the meaning of science and its relation to nearly everything else. Scientists are taught to privilege "intellectual" concerns (and one's growth in them) over economic, emotional, and social ones, and intellectual growth (usually) stands for growth in general to them. Scientists also often associate "play" with creativity and innovation in their work as a positive characteristic. Their image of *science* as a unilinear, infinite, progressive movement of knowledge is modeled in their own *self*-image as continually maturing or developing personae. Similarly, managers construct their self-images in their own pattern of completed adulthood, which informs the model of their finite, finishable, productive reality of work. Scientists then cast a jaundiced eye on this image, and consider managers, not as "finished adults," but as cases of arrested or frozen development. Recall their comments about the "narrow" managerial focus on economics. To scientists, the managerial interpretation of self has stopped the "natural" process of continuous growth. Thus, they can accuse managers of a generic deficiency in the development of the "sightedness": they can only appear as short-sighted to a scientist.

No wonder scientists and managers find it hard to cooperate with and understand each other. In their grossest caricatures of each other, "the complete adult realist managers, in their struggles with immediate economic necessity, must contend with immature scientist-dreamers; while from the other side of the table, the far-sighted progressive scientists must protect their work—the bases of the firm's wealth—from myopic and developmentally retarded managers!" Moreover, each group brings to their encounter the self-assurance of membership in a very high-status group in American society, be it the academic universe or the business one; and in both their cultures of apprenticeship, this brash self-assurance is nearly a *sine qua non* of attaining higher levels of professional success.[25] An example from an argument between two principals from these communities illustrates how these images appear in practice. They are not always so extreme as caricature, but a public forum requires a certain politeness to mask or soften disagreements.

A Sample Argument

We can see these characterizations of self and the other in the following interchange between the CEO of a large company involved in biotechnology and a senior biology professor. The context is a small discussion, where the biologist has brought the executive on campus to meet with a group of postdocs and students. At this point, they are talking about the company's relationships with academic biologists, and how to expand them. The CEO has spent several minutes talking about what his company wants from university scientists—in short, "trained people . . . and new ideas." He then suggests five kinds of relationship: (1) consulting on "specific questions with *specific* expertise"; (2) "specific project support, for instance, for clinical studies; these are where we have short-term, specific goals such as proving the safety and efficacy of a specific drug"; (3) project support in academic labs "which grows out of [the specific needs addressed in] the first two categories, and which depends a lot on the personal relationships developed there"; (4) "unrestricted but focused" funding of university research, "like Monsanto at Harvard or Johnson & Johnson at Scripps"; and (5) "focused philanthropic giving" for endowed chairs and fellowships. Noting that the last two categories are trivial percentages of overall research budgets in industry or the university, he goes on to say that "contacts [are] *not* more extensive because of fundamental differences. You academics work in unstructured ways, while industry wants a detailed structure." The CEO also criticizes a "Boston academic perspective"—a reluctance to assign exclusive patent rights to the company that sponsors research, saying that this reluctance is at odds with "the real world, where companies will go wherever institutions will grant those [exclusive development] rights."

At this point, the senior biologist takes exception to this argument, and offers a different interpretation of the same process, that "short-term [motivated] choices among schools which offer 'better' deals may lead to long-term debilitation of the university's value to industry." He argues that firms should "invest" in the *long*-term value of the university's science-generating (and scien*tist*-generating) capability by supporting the last two kinds of open-ended relationships (items 4 and 5 above). In making his points, the biologist even uses "economic" metaphors like "ROI" (return on investment) to characterize this process.

The curt response of the CEO is that scientists can't "plan discovery"; so their main engagement with (his) company should be over concrete, specific projects.

Several points can be made about this interchange. Notice, in the CEO's first talk, the emphasis on short-term "deliverables," specific projects—that is, bounded, closed-ended ones [items 1–3]—rather than long-term, open-ended ones (items 4 and 5). Even the "more open-ended" relationships 4 and 5) stress a *focus.* The biologist's rejoinder stresses "long-term value." The CEO characterizes the university biologists directly as working in "unstructured ways"—read, open-ended—whereas industry wants closure. The "real world" for the CEO is a short-hand for specific, bounded projects, whose funding is driven (or sponsored) by short-term economic considerations—that is, "Which institution gives the best licensing deal *today*?"

Finally, the biologist tries to use the stereotypical arguments of business—in a rhetoric of "investment" and "ROI"—to turn the argument around to his own favor about what to fund. The biologist *expects* the CEO to argue in these terms, so he tries to turn that rhetoric to his own advantage. The CEO, in return, draws on the common image of scientific discovery as "unknown," and thus "unplannable." Each expects the other to understand and manipulate a set of more-or-less stereotypical images; and each is familiar with the other's repertoire, at least enough to come up with recognizably appropriate images, appropriate, that is, to the context of argument. In an ironic reversal of characterizations, each draws on the commonly expressed self-images of the other party to argue why and where money should or should not be spent. They use and exchange these images as tools of argument; but, here, the *process* does not bring them any closer to agreement.

STATUS PEAKS AND POWER DIFFERENCES

Having described what distinguishes the culture-worlds of these particular scientists and managers from each other, I want to remind us of some important congruences. Both groups, in their native habitats, are accustomed to being at the top of their respective status hierarchies. Presidents, CEO's, and entrepreneurs are in the top ranks of the high-visibility, high-technology business world. Likewise, the university faculty involved in biotechnology companies (and most of the postdocs) come from elite university departments. Just as the biotechnology/genetic engineering industry has been in the recent limelight of prestige in the fast-moving high-tech start-up world, so too are molecular biology and immunology disciplines "on a roll"—at the cutting edge of major scientific discoveries, well funded, and in the forefront of public discussion of science. Each group is also used to running its own show. The scientists, as laboratory heads or faculty researchers in universities, tend to direct their work in an environment that is relatively free of immediate administrative constraints. Furthermore, they image managerial or planning constraints as "external" ("social") and virtually antithetical to the practice of science. Executives, though, in the company environment, for their part expect to manage the scientists' work too. Managerial use of detailed, short-term plans is also a tool for achieving control, both of costs (specifically) and of company activity in general. (See also Chapter 3, above.) Each group has its own ways of building its authoritative voice—often at the expense of the other—but each group also insists on the primacy of its own authority.

It must be kept in mind here that managers also have the power to direct or manage (or hire and fire) most of the bench scientists in companies.[26] The dual model of culture-worlds I have sketched indicates a separation or difference. This difference is certainly affected in practice by the asymmetry of power in the organization. The power to manage or direct, however, cannot always easily be exercised in a straightforward, bare fashion.[27] Scientists can wield the weapon of their technical expertise, at least to some degree, to counter managerial pressures. The scientists control a kind of "secret knowledge," largely inaccessible to non-scientist managers, and scientists can use this expertise as a rhetorical tool or weapon to fend off forays by management into the details of laboratory practice (including expenditures). Likewise, of course, managers can use their detailed knowledge of networks and procedures—for instance, in finance and accounting—to isolate a realm of technical expertise of their own in this contest.

BUILDING DIFFERENT CULTURE-WORLDS

The biologists and managers recognize and emphasize their perceived differences from one another, and they manipulate these images of contrast with facility. At the same time, from the sample of argument above, we see that they have begun to learn to use each other's rhetorical styles and arguments. How can we begin to account, then, for some of the tenacity with which each group also maintains its own repertoire of images? Images of self and work as well as the temporal patterns for planning work seem deeply ingrained. Even two or three years into a firm's life, scientists and managers report the same kinds of conflicts drawn in the same rhetorical tropes that they reported at their firm's nascence.

One place to look for the geneses of these differences is in the education and socialization of managers and scientists, particularly in the habits they cultivate—their cultural patterns—for problem solving. For scientists and managers both, there is a long formal period of training and apprenticeship in which cultural patterns are taught, inculcated, and valued.

Scientists from top universities go through a very intense and taxing initiation process to get their PhDs. Potential biologists often begin to concentrate or specialize in science studies as early as high school. By college, they begin a series of time-intensive course activities (including labs) that inexorably tend to push other subjects like the arts, humanities, and social sciences (or even other natural sciences) out of their schedules. This involves years of "background" course work in the basics of their science, materials to be silently absorbed as an undergraduate and young graduate student. Then, as a PhD candidate, the scientist must design and complete a relatively solitary piece of research. This last process may take three to five years (or more) of 10- to 14-hour days in the labs. Only after completion of the PhD, as a post-doctoral researcher, does a scientist begin to work in a more integrated research group on a common project, where work is imaged as cooperative (and faster).

To an anthropologist like myself, who has done fieldwork in East European and in West African agricultural communities, the single-minded focus, intensity, dedication, and isolation from other arenas of social life of many young novice molecular biologists seems extraordinary. On the other hand, having worked as a technician for four years in similar university wet biology labs (1970–1974), I can attest to the impression of "naturalness" such a nearly "total" environment creates as a lived-in experience for its native biologists.[28]

For the managers, my observations of culture and socialization in the formal education processes at Stanford's, MIT's, and Harvard's MBA (Master of Business Administration) programs point to a different picture. A two-year MBA training is no less intense, but it cultivates a different style of engagement with the world. MBA students are taught—almost forced sometimes—to take a combative, self-assertative stance in public forums. They are called upon in classes to "recite" and argue with each other and with their teachers. By contrast, the early training of the scientists emphasizes "learning the material" in a more passive fashion. For science students, "the material" is construed as "facts" that are largely unproblematic at the early stages of one's career. This attitude fosters learning as the progressive accumulation of information at least through an undergraduate degree, and through many formal graduate courses.

Contrasting notes from my observations at a graduate seminar in biology at MIT with a Stanford MBA class is illustrative:

> It is the first meeting of the [biology] class. About forty students sit in rows facing a blackboard, talking quietly among themselves. The professor arrives, and talk stops immediately. There is a rustle of paper as students ready their pens and note-pads. Without a word, the teacher strides to the blackboard, writes [her or his] name, the course *number* (but rarely the title), meeting times, and office hours on the board. The first words recite these same inscriptions. The students diligently copy this information onto their pads. Dismissing these "administrative details," the lecturer then launches into an introduction to the course material.[29] Aside from a few tentative questions—usually from student to teacher, talk is dominated by the professor. Students write extensive notes, seldom speaking; and they always address only the teacher (not each other), if they speak up at all.

Their primary task is to accumulate "background" information—a foundation in the field's basics. It is only in more advanced, specialty seminars that students are encouraged to participate in discussions, often prompted by their teachers. The self-assurance seen in top molecular biologists is only just beginning to be cultivated in students at the end of their graduate careers, and it is progressively more encouraged in postdocs.

My observations from attending business school classes reveal a very different pedagogic ambience:

> The classroom is built as an amphitheater. There are tiers of seats with writing tables in front of them, drawn in a semicircle facing the professors's "stage." As students arrive, they place their foot-long name-plates in the slots before their seats. Then, they move about and congregate in small groups, a buzz of talk humming on about their work assignments or topics of the course.

Students are encouraged to collaborate in most courses. In fact, they are required to do so; and the social sorting and jockeying over who will work with whom is an arena of intense competition. Students may know the reputations of their classmates from guides even before they arrive, and those who have special expertise or valued qualities are intensely recruited by peers who want to share these resources in preparing reports and presentations. While group work is promoted, there is also a tension between a need to demonstrate individual accomplishement versus a desire to have one's "team" excel.

> The teacher's role seems almost combative: groups and individuals are challenged to answer pre-set dilemmas or problems, and the teacher contrasts one group's ideas with those of another to stimulate argument.[30] Students have a very forthright, self-assertive, almost brash style of presenting ideas; and there are kudos for the "delivery," as an integral part of the solution. It is not enough to simply have the "facts" to make a good argument, they have to be presented with *élan!*

Part of that compelling presentation is the self-assurance that conveys strong conviction in the ideas one presents. This style of *self*-presentation is crucial to an executive's demeanor, too, both in making important managerial, economic, and personnel choices within the firm and in dealing with present and potential investors outside it.

The business school case method of pedagogy, however, also usually limits students to consideration of the issues presented in the written "case" handout. The kinds of questions that are relevant are usually circumscribed or pre-determined by the parameters laid out in the written materials. Students are thus taught to examine, analyze, and make decisions carefully about situations whose boundaries and parameters are largely defined, and then they skillfully argue about their choices. Speed is also of the essence; students commonly have only a night or two to analyze two or three written cases of 20 or more pages. I am not proposing that students do not learn to think cleverly and creatively, or that they are not given to careful analysis. Rather, I would suggest that they cultivate through this pedagogy a habit of working best within the confines of a problem whose parameters and limits, like the cases, are finite and (preferably) knowable. It is no surprise that any decision-maker might like to know the bounds of a problem; what I argue is that business schools are especially good at cultivating the complex treatment of bounded issues—seeking closure. Managers socialized in this environment are thus more likely to seek boundaries to problems and issues, so that they can work toward closure on them in the fashions to which they have been accustomed (and trained). MBA students—and the managers they become—learn to use "facts" as tools of contention in the context of problem solving and decision making; and they prefer to bound their problems, and then attack them in measured stages. Seeking these boundaries and parameters in the stages and milestones of company plans then runs headlong up against an equally well-ingrained but *different* cultural style of problem-solving work among the biologists.

Scientists and their apprentices have a different set of approaches to the structure of problems, and these change over the course of a scientist's career. One way to characterize this change is as a general movement away from bounded problems, designed and presented as pedagogic tools, toward an examination of nature that holds unlimited surprises for the scientific investigator. One aim of a developing career is movement toward issues that cross the boundaries of present knowledge. At the beginnings of a career, the majority of "problems" presented in introductory textbooks through the graduate level are designed to get students to apply some particular set of knowledge to a relatively circumscribed situation. These are the "thought" problems that can be assigned as homework or tests and the laboratory assignments that lead students through a fixed protocol. Just as important as teaching students how to think and reason at this stage is teaching them to do careful step-by-step work and to keep scrupulously accurate records of it.[31]

As students enter advanced graduate seminars and begin to look for PhD thesis topics, the character of their work begins to change. They are introduced more and more to the contestible assertions of their disciplinary specialty, and they are encouraged to take positions on some side of an argument. As they look for a thesis topic, they may choose to enter a controversy or develop research in an area that promises (usually) to expand the margins of knowledge about some currently interesting or puzzling natural system or substance.

The key to this shift is that the parameters of the problem are usually imaged as being established by the *natural* system—not by the definitions of the investigators (or teachers). This is an indication of a principle frequently invoked by scientists that the "answers" to their questions are governed by "nature," and that their investigations are a process of seeking knowledge about a system that is virtually unbounded. The "problems" or issues are posed by nature, and thus the construction of their parameters and boundaries depends upon exigencies external to the scientists' ability to investigate them. "Nature" is in control, in an open-ended system, and scientists are seekers. This is *not* an attractive argument for business executives.

It is, however, an occasioned statement of the case, *designed* to accentuate their conflicting approaches. Biologists do not always emphasize the same rhetorical properties of their work. By contrast, when students seek a thesis topic, for instance, their mentors usually intervene to assist in constructing what is considered a "do-able" project. That is, the parameters of some issue—generally one well defined in the social context of contemporary scientific work in the field—must be clear enough to constitute a relatively closed system, a question that may be posed and answered (either positively or negatively). Responsible mentors try to help their students find a problem (or select one for the student) that fits a balance between advancing into some unknown region of knowledge, so that the work is not "trivial," and yet staying near enough to the centers of accepted work that the student will neither become mired in too complex a problem nor be stymied by a lack of supportive investigation in the same scientific "neighborhood." The questions should be relatively small, but not trivial, and relatively well embedded in a universe of correlative scientific discourse—that is, not isolated. For the ambitious student, the trick then becomes to push the boundaries of this envelope of mentorial attention outward. This means finding as innovative and interesting a problem as possible, without biting off more than one can chew. A successful PhD project requires both good analytical work and careful experimental practice; but a superlative recommendation from one's mentor may require the student to demonstrate just a trifle of disregard for the careful approach in accomplishing a successful attack on a problem.

These shades of negotiated difference are similiar to a cultural distinction between two native descriptions of what scientific activity is. One kind of description stresses the Baconian model of hypothesis testing: the rigorous examination of a question or issue according to a program or experiment that will reveal or demonstrate the truth value of the hypothesis. This is a planned and bounded image of practice. An alternative description stresses the unknown character of some phenomenon to be investigated, and emphasizes hypothesis formation and discard in an active, searching process. The latter image seizes upon the open-ended activity of seeking in an unfixed or unknown region, and it evokes the paradigm of unbounded nature as a model for the parameters of practice.

In the decision making of biotechnology firms over research and development choices and directions, as in the negotiations over a student's dissertation topic, either of these two contrasting images of scientific work may be alternately chosen as the bases for some particular argument. Biologists with whom I spoke could switch rather freely between the two images for different purposes of explanation or clarification in our discussions, and it is their artful management of these differing rhetorical strategies of argument that allows them to accent either the "empiricist" or the "contingent" qualities of their work (Gilbert and Mulkay, 1982). This argumentative flexibility can become an important rhetorical tool in biologists' interactions with managers in their firms.

There is a further difference in the formal pedagogy of managers and scientists that becomes central to their grounds for argument. The training of managers and their subsequent practical experience teaches them to place a high, if not central, value on economic productivity and the generation of profit. This, to business people, is an expression of the most serious concern with the "real world", and this particular "real world" is considered, not just as one choice among many, but as *the* most important and pervasive activity. Much of human activity (particularly *in re* organizations) can, for them, be reduced to economic motives; and company policies and plans must conform to this reality. While scientists, on the other hand, are certainly not unaware of economics, their fundamental views of reality include a primary focus on "natural" systems—the systems that they investigate professionally. Their training focuses on the value of scientific study of this reality and on the importance of the products of their work: the generation of "knowledge" (Knorr-Cetina, 1981) and "scientific facts" (Latour and Woolgar, 1979). They often see these aims as different from or opposed to economic motives, and they usually consider economics and business as extrinsic to their science proper. While they acknowledge the "influence" of society, economics, and organizational issues on their ability to accomplish their work, they still see the scientific "content" of their praxis as drawn by nature and as separate from its "social" context.[32]

DISCUSSION AND CONCLUSIONS

Conflict, Communication, and Difference

The biologists and managers have different socialization and training, goals and values, styles and processes of work, and ways of understanding themselves and their professional context. The opportunities for conflict are exacerbated in most firms by a consistent segregation of physical space, functions, and communications between managers and scientists. Firms are organized into two separate hierarchies. Managerial staff, finance and control, marketing, and operations and production, if they exist report directly to the CEO. The much larger scientific and technical staff report to lab or section heads (scientists), and they to a director of research (perhaps an executive vice-president), who then reports to the CEO. Communications flow up, down, and around within each group, but managers and scientists rarely interact except at the highest levels of the organization. They are spatially segregated—managers on a "mahogany row" or sometimes even in a separate building, and buffered by gate-keeping support staff. They maintain their different and separate identities through contrasting styles of appearance, dress, and the decoration of their native office or laboratory habitats.

Scientists and managers commonly complain of being misunderstood by each other, and companies often hire a "translator" to bridge this perceived communications gap. But the complaints continue. Yet both groups must somehow collaborate toward common goals of product definition and development, or research direction. Managerial decisions and scientific research ones depend intimately on each other, and yet the oral histories of firms are rife with tales of marketing groups forcing the cut-off of an *X*-month project without consulting the research workers. The sudden layoff of 15 percent of the workforce in autumn of 1984 at Biogen (Cambridge/Geneva) and at Genex (Rockville, Md.) or the termination of the "new ventures" group at Cetus three years before are just a few drastic examples. Because of the power asymmetry between managers and scientists, "decisions" are ultimately in the hands of the executive group, however much or little they consult their scientific advisors and staff. The paucity of ongoing, overlapping interaction and common understanding across functions in the organization, however, seems to contribute to "decisions" being the "cutting off" that the etymology of their Latin root implies.[33]

Conflict per se, though, is not necessarily all bad. There is a significant management literature to suggest that conflict can stimulate a healthy interaction of ideas, avoid "group-think" agreement, or foster innovative solutions. Others also contend that fostering internal competition and conflict, even of a highly contentious sort, will, in a kind of metaphoric Darwinism, select for the "best" solution.

Some form of "stylized" or "ritualized" conflict may become embedded in the cultural patterns of the organization, just as it is reported in some other high-technology industries. Dyer (1982), for instance, reports a style of "beating on ideas" as a way of discovering "truth through conflict" in a high-tech microelectronics manufacturing company. The context in which conflict occurs, however, is critical. Lively debate, intense planning interaction, and a highly interactive overlap between functional groups at an early planning stage is becoming recognized as an appropriate strategy for managing high-technology environments. A key to this strategy, however, is not fighting over turf and resources, but rather cooperative attempts by different groups for the mutual support of a common project. Another key is timing: do the up-front planning and disagree *before* large resources are allocated; don't let change and indecision drag on to disrupt schedules and reopen old arguments.

By contrast, in the early 1980s ongoing, unresolved disagreement seemed to plague many biotechnology startups, and communication was often meager. Managers and scientists repeatedly emphasized the "otherness" of their counterparts. Some managers left the scientists "on principal" to do their thing, not involving management much until issues arose of how to market project results. This practice follows the common managerial mythology that scientists must be left alone—and want to be left alone—to do their creative work. Bailyn's research on autonomy and professional satisfaction in industrial science, however, debunks this myth. Company scientists in her studies responded best to corporate strategic goal setting for focusing overall directions, while they were best given operational autonomy to do their day-to-day work of finding their specific pathways to those goals (Bailyn, 1984). This calls for measured managerial interaction with scientific planning, not avoidance. Some other biotech managers engaged in continuous, indecisive arguments with either R & D directors or their scientific advisory boards over which projects to pursue. Many companies seemed to run off down every "promising" path, unable to set priorities in research when results seemed so hard to project in the future. This strategy rapidly dissipates resources, and prolongs the ongoing argument over what is to be done. Each approach leads to an eventual "cut-off," where the parties are nearly bound to be arguing over the distribution of scarce resources. These are the arguments reported to me from the field; and, while they may have diminished in force and number, they have not disappeared. By the mid-1980's, biotechnology firms were gradually finding different ways to accommodate and manage their disparate professional cultures, but the rocky relations of the early years of some companies and the industry were not entirely smoothed.

Furthermore, the cultural differences between these professional groups have certainly not disappeared. But cultural difference alone does not necessarily entail conflict. The conflicts that arose also grew partly from the hubris of major actors on each side, a hubris fostered by cultures of individual, competitive excellence within both groups. The conflicts were also anchored in arguments over the distribution of material resources—space, equipment, salaries, and perks—as well as the social resources for enhancing status and reputations. Stakes were high, egos were large, and the groups were initially unfamiliar with each other. As managers and biologists have lived and worked together for five or six years, they have in some cases learned to treat difference as an issue or even an asset to be managed, rather than as a force to be reckoned with or overcome. The successes or failures of young companies are not the subject of this chapter, though, but rather the delineation of those differences in professional cultures that *make* a difference to managers and scientists. These are the differences that engendered early conflicts and that persist as important characteristics of organizational life.

PATTERNS OF CULTURE: HOMOLOGIES ACROSS CONTEXTS

Time is the key symbol that focuses the cultural patterns of pedagogy, apprenticeship, and problem solving across different contexts of social action and experience. These contexts include the development of professional careers and personae, the core practices of professional work and its planning, and the character of fundamental realities—the nature of nature and of business—within which each group does its work. Time is the lens through which these diverse contexts cohere into a characteristic cultural configuration, a pattern of homologies across and among contexts of action and experience. These patterned homologies constitute a fabric of meaning for each group, embedded in its historical experience and daily activities, which engenders the group's distinguishing character as a cultural variant—of biologists or of executives. This cultural specificity lies not just in their ideas, norms, values, or images of the way things are, but equally in their concrete activities—the technical and professional work through which each group constitutes itself as a community. Career-time models and project-planning styles are guides for *doing* things: cultural patterns whose images of the world are compelling to their subjects precisely because those subjects enact, live through, experience, and validate these models directly in the long course of building their careers and making plans.

For managers, equating the "real" and the "economic" conflates a vast array of social and "technical" interactions into a system for producing tangible products—results with an end. These are bounded projects, and I have argued that the pedagogy of business schools is a training system for elaborate analysis of bounded problems. The "case method" of teaching, pioneered at the Harvard Business School, is the exemplar of this Socratic method of presenting "management dilemmas" for discussion. Training in some of the more "rationalized" specializations like finance, control, accounting, decision analysis, and operations research also often focuses on targeted, small-scale problems. In business practice, a problem-oriented focus and a world view that privileges the reality of traditional "economics" tend to direct managers' attention to the nearer-term issues; and this is partly because an urgent "economic reality" of quarterly reports and the next round of funding constantly impinges itself on plans. Plans are a necessity, and the shorter the better. A staged, analytical sequence directed to closure on a finite goal fits the problem-solving models taught to business students, and its "completeness" resonates with the models of the responsible adulthood that inform a business career. Many top executives in biotechnology start-ups come from venture capital and finance worlds in which these characteristics are strongly cultivated. As culturally constructed personae, managers—even entrepreneurial managers—present themselves to the world as "completed" adults who usually work in a world of bounded, finite projects, whose progress can be tracked in "timed" stages to completion. Their cultural models of time in careers, as selves, and for professional work share a homology based on "boundedness," "completion," or "closure" across all three contexts of action and experience.

For the biologists, the overarching "reality" lies in nature, and, to them, this natural reality is conveniently separated from, and prior to any social reality. The distance of nature from the social realm allows it to be used as a valuable resource in argument. The "real" answer is just beyond the present limits of knowledge, always drawing the investigator further into an unknown. Just as the object of scientific work is the unknown—what lies beyond the bounds—just so is a scientific career a continuous search for these ever-elusive ends; and the intellectual development of a scientist is likewise an unending, unbounded process of growth. The scientists' model of nature as an open-ended infinite system matches their cultural pattern of time. This open-ended time is a model for practice in professional work as well as a model for continuously developing scientific careers and selves.

Finally, for the cultural study of organizations, this research argues that differences *within* organizations can be more powerful shapers of organizational life than some blanket "corporate culture." The cultural variants peculiar to professional groups within these biotechnology firms contrast strongly enough to compete with any overall company ethos. In addition, understanding these differences requires examination of each of the professional groups' technical work; that is, technical professional work is culturally shaped. In the biotechnology case, addressing the interaction between two groups in their joint planning work served as a tool for contrasting their professional culture variants. Attention to technical detail as well as the holistic interpretation of cultural pattern across multiple contexts is common professional practice in anthropology. Only the "natives" in this study are uncommon ethnographic subjects, though they are our close neighbors and familiar in many respects. That so fundamental a category as time can be so different for each of these groups gives the lie to any facile assumption that "Western culture" is a rational unity against which foreign "others" are implicitly compared. Both scientists and executives can and do lay claim to Western "rationality," but the ways in which they understand and exercise it reveal a fundamental diversity. This diversity is strikingly apparent in their cultural constructions of time.

NOTES

Acknowledgment: First, I owe a great debt of thanks to the many busy company executives, scientists, students, and staff from some two dozen firms and several academic institutions who have given freely of their time in this project. The interpretations I present are partly theirs and partly mine—a collaborative effort at understanding issues of mutual concern. If they miss the particular details of their own circumstances, I apologize for the synthetic brevity of this account.

MIT's Science, Technology, and Society (STS) Program, where I held an Exxon Fellowship, and its Anthropology and Archeology Program have provided material and intellectual support for this project. The STS Workshop on Engineering and Scientific Practice, organized by Larry Bucciarelli and Sharon Traweek, provided valuable criticism and insights; and I thank those two along with Steve Barley, Antonio Botelho, Jon Gulowsen, Martin Krieger, Eric Livingston, Laura Nader, Ed robbins, Wolf Schäfer, Edgar Schein, Dennis Sebian, Knut Sørensen, Sherry Turkle, and Steve Woolgar for their contributions. Several other faculty from the Sloan School of Management have also made helpful suggestions and criticisms, and I especially thank Tom Allen, Mel Horwitch, and John Van Maanen. Thanks, too, to Joanne Martin (Stanford, Business School) and Sylvia Yanagisako (Stanford, Anthropology), who assisted my research in earlier stages, and to Jon Anderson (Catholic University, Anthropology), Sukhendu B. Dev (Battelle Memorial Institute), Eugene A. Hammel (UC-Berkeley, Anthropology), Sandra Panem (U.S. Environmental Protection Agency), and George Rathmann (Amgen).

The work was first presented in a session on the Ethnography of Science and Technology at the 1983 American Anthropological Association meetings, and I thank the other panelists. Jeanette Blomberg, Carol MacLennan, Daniel Segal, Sharon Traweek, and discussants Mary Douglas and Stephen Toulmin for their critical commentaries. I also thank Catholic University, the University of Illinois, Michigan Technological University, and Northeastern University for opportunities to present this material and receive valuable feedback in colloquia.

Finally, special thanks are due to Malcolm Gefter (MIT, Biology), Anna Hargreaves (Lotus Development Corp.), Katherine Hughes (Sage Foundation), Evelyn Fox Keller (Northeastern University), Sharon Traweek (Rice University), and Charles Weiner (MIT, STS) for many hours of rich discussions on this work and to Anna Hargreaves, Dave Pearson, and Edgar Schein for their careful critical readings of this text.

1. My own research has encompassed a variety of tactics and methods. They include extensive interviews with a variety of academic and company personnel from all levels of the organizational and professional hierarchies (from CEOs and senior biology faculty to technicians and students), although I have generally concentrated on the more powerful actors. I have also visited labs and companies for up to a day at a time, several sites more than once, and have been led around facilities by a variety of scientists and managers who have given generously of their time to assist me. In addition, there is a massive literature on the industry and on the recombinant DNA controversy; and many of the principals participate (and argue) in public fora, where I have occasionally taped their interchanges. The MIT Archive has an excellent collection of interviews from the mid-1970's with principals in the rDNA debates, including some industrial executives and many scientists who now participate in companies.

2. "Cloning a cell line" refers to a process of selecting a single parent cell (or genetically identical set of cells), usually ones that have been genetically modified to enable this selection, and encouraging it to reproduce itself asexually as a "colony" of identical cells. Of particular relevance to modern biotechnology is the development of hybridomas—novel cells that are made by fusing a myeloma cell, which divides rapidly in culture, and a lymphocyte, a cell that produces antibodies. The resulting hybridoma produces *monoclonal* antibodies, homogeneous antibodies that recognize only one (chemical) antigen. The enabling work was done by Cesar Milstein and Georges Köhler in 1975, for which they won the 1984 Nobel prize in medicine. See Milstein (1980) for a summary account of the technique. Cloning technologies become crucial for a parallel development to gene splicing in the new industrial biotechnology: that is the commercial development of monoclonal antibodies, and especially their applications to diagnostic testing.

3. I attended and "observed" various classes, lectures, and informal social activities at the Stanford Business School, MIT's Sloan School of Management, and the Harvard Business School. Several scholarly and popular books and articles have discussed the ethos of these schools (and their differences); and Peter Cohen's first-hand account of student socialization at Harvard Business School is a particularly rich description of the interaction styles that are fostered there (Cohen, 1973).

4. Large petrochemical and pharmaceutical companies have also lately become involved in the "new biotechnology." These include DuPont, Monsanto, and Dow Chemical, and Merck, Eli Lilly, Schering-Plough, and Johnson & Johnson, among others in the United States and abroad. Their involvements include a wide range of strategies from in-house R & D commitments to various contractual and equity relationships with universities, individual academics, and small start-up companies. Only these latter entrepreneurial ventures are the focus of this chapter. See Friar and Horwitch (1984) and Horwitch and Sakakibara (1983) for a description of the range and complexity of these relationships at all scale levels.

5. In simplified terms, DNA can be seen as a string or sequence of "bytes" of information. Each byte is three "bits," that is, three adjacent base-pairs in the long strands of the twisted DNA molecule, which Watson and Crick described as the "double helix" (1953; Watson, 1968). Some unique series of these triplets defines a "gene," a unit of hereditary information that, under the proper circumstances, expresses some trait in the organism.

This description mirrors the "central dogma" (Crick, 1958, 1970), the view widely held until the mid-1970's in molecular biology, that "DNA produces RNA (ribonucleic acid), which produces protein," and that this is a unidirectional process. Since the late 1970's, with a growing interest in issues of gene expression and regulation and of cell and organismic development, new research has revealed the need to supplant this simple model. The mechanistic image of genetic control has been superseded by an acknowledgment of the complexity of intra-cellular and environmental interactions with DNA. While recognition of this complexity was being addressed in classical cytogenetics during the same years as the rapid development of molecular genetics (1950 onwards), the latter field dominated the limelight. It was not until the late 1970's, for instance, when molecular biologists discovered "jumping genes," that they were compelled to acknowledge that the same phenomenon had been described as "transposable genetic elements" by the great maize geneticist Barbara McClintock almost 30 years before. The differences in styles of work and the lack of interaction between these two branches of genetics are described in eloquent detail in Evelyn Fox Keller's work on McClintock and genetics, *A Feeling for the Organism* (1983). In the early 1980's, concerns for the complexities of living systems are being addressed by molecular biologists, but still largely within their own mechanistic and reductionist paradigms. See expecially Keller (1983), pp. 180–81, 192–195. For a detailed history of the development of molecular biology, see Judson, *The Eighth Day of Creation* (1979). As late as the publication of Judson's work, however, he still speaks of the "central dogma" as a recognized principal (p. 612). See also Fuerst (1982) for a discussion of the centrality of reductionism in the development and heuristic models of molecular biology.

6. Documents from the history of the recombinant DNA controversy and the birth of genetic engineering are collected, with commentaries, in Watson and Tooze (1981). Developments in the molecular biology of DNA through the early 1980's are presented in graphic form, for the educated lay person in Rosenfeld et al. (1983), and a simpler picture of the mechanics of molecular genetics may be found in Gonick and Wheelis (1983). For a more detailed discussion of techniques in genetic manipulation, see Old and Primrose (1980). Cherfas (1982) also details the historical development of the enabling science and technologies of rDNA, and goes on to discuss their commercial application.

7. See Cherfas (1982), pp. 142–143, 155–167, for a detailed discussion of the enabling scientific work leading to commercial "genetically engineered" insulin.

8. The OTA Report (1984) is an excellent source for an optimistic appraisal of prospects for the new biotechnology industry.

9. See Sheldon Krimsky's *Genetic Alchemy: The Social History of the Recombinant DNA Controversy* (1982) for a detailed

treatment of the development of the issues of "biohazards" as the primary focus of debate around rDNA.

10. Sandra Panem (1984) discusses this dynamic in detail for the commercial development of interferons.

11. Many industry publications credit Japanese industry with a more advanced fermentation technology sector than the United States has (e.g., Hochhauser, 1983; OTA Report, 1984). New techniques in eucharyotic cell culture may, however, supplant large-scale batch and continuous fermentation of bacteria and yeast, as new hosts are found for recombinant genes.

12. Joan Fujimura has done pioneering work on the social construction of "doable" research projects in the molecular biology of cancer, the subject of her PhD thesis at UC-Berkeley (Sociology) and the Tremont Institute of San Francisco. See also Fujimura (1987).

13. This opportunity or promise rapidly waned, as the exigencies of economic planning forced companies to cut back on basic research, as they struggled to develop marketable products. For instance, even before the change of CEO at Cetus Corp. in early 1982 from the visionary founder Ron Cape to the industrially experienced Robert Fildes, their "state-of-the-art" new ventures scientific group was disbanded (third quarter 1981), several of its scientists were laid off (about 40 staff altogether), and others were redistributed among other working groups.

14. I have discussed a similar phenomenon for Central Australian aborigines, in the context of their movement between (loosely speaking) "religious" and secular" worlds (Dubinskas and Traweek, 1984), and plan to discuss the multiple constitution of "selfhood" for Western professionals in more detail in future work.

15. Many industry observers and firm members express skepticism about the general strategy of importing R & D mangers from big companies to tiny start-ups. They argue that the managerial work of administering a large R & D budget, which is a relatively fixed proportion (usually less than 5 percent) of a massive industrial budget, in a relatively stable environment from year to year, is drastically different from juggling a similar amount of money in a high-technology start-up. Issues of cultural style as well as the practical experience of differing man-

agement problems often arise. Small firms are more volatile in structure, economics are dealt with on a face-to-face basis with CEOs, staff are less formal in attire and interaction, and decisions are not made in such a hierarchical or bureaucratic fashion as in large firms and their R & D divisions.

16. Until perhaps 1982, few large companies were willing to invest in in-house research efforts with gene splicing, and the majority of them still prefer to establish joint research ventures or projects with small genetic engineering companies rather than build house staffs. Exceptions include Monsanto, with its large commitment at St. Louis (still in conjunction with Washington University), and DuPont, which in the autumn of 1984 announced the establishment of a new research center in Wilmington that would include biotechnology work.

17. I try to choose my style of dress to be relatively appropriate to the environment I am visiting in a firm. Suits are *de rigueur* for managers, but a sport coat and tie will do for anyone short of research director. There is a difficult guess for meeting academics on campus; senior professors sometimes inhabit their labs in blue jeans and their board rooms in business suits. I usually overdress for arrival and "loosen up" later, if it seems appropriate.

18. See also Knorr-Cetina and Mulkay (1983) for additional works in and comments on this genre of studies.

19. I discuss some early dilemmas of conducting participant-observation field research in a managerial environment in Dubinskas (1984).

20. Allen's (1966) work was intended to speak to different issues from those for which I have chosen it as an illustration. His interest in this work was to look at processes of decision making in R & D environments to see if generalizations could be made about parameters of adoption of particular choices, without considering the technical contents of a choice.

21. The permeation of the professional culture of molecular biology by the ethos of physics is the subject of active and lively discussion and investigation. See, e.g., Keller (1984, "The Heritage of Physics in the Transformation of Biology," and 1983), Dev (n.d.), who interviewed many Harvard and MIT molecular biologists, and Abir-Am (1982), whose paper introduces a large bibliography on the subject.

22. In contrast to Traweek's focus in Chapter 2, I am concerned with temporal domains *all* of which would be in the realm of non-relativistic time for the physicists. While Traweek is centrally concerned with how machines mediate contradictions between relativistic or immutable time on the one hand and non-relativistic or ephermeral time on the other, this chapter discusses patterned relations of similarity among domains that are all under the same (non-relativistic) rubric for the phsycists.

23. This applied scientific work, based on a "mature paradigm," may be an indication that molecular biology (or some part of it) is the kind of system that Schäfer (1983) and his contributors describe in their work on "finalization" in science.

24. Metaphors from Tom Wolfe's *The Right Stuff* (1979) and its subsequent movie version by the same name have become commonplace in the language of the scientific and technical communities, at least around Cambridge and Silicon Valley. "Punching the envelope" is an expression used by test pilots for flying their planes beyond the "accepted" limits of their capabilities. Testing limits and breaking rules is a characteristic pattern in American culture that Gregory Bateson describes in his work on double-bind theory (1972).

25. Making it to a senior post-doc or junior faculty level in molecular biology sems to require cultivating a similar forthright and self-assertive social presentation to the one the business school students develop at an earlier stage in their careers. Some scientists even suggest that a certain arrogance of style is an asset in projecting the strength of convictions in one's own capabilities and ideas.

26. It would be a rare working company scientist who controlled enough equity in a firm to prevent its managers from removing him or her. Walter Gilbert's departure from active management at Biogen may be a case in point of the eventual power of managerial and investment concerns over even the most dynamic and prestigious scientist-entrepreneurs.

27. The 1981 termination of many bench scientists at Cetus and the layoff of 13 percent (or more) of Biogen's work force in November 1984 (Fox, 1984) may give the lie to this observation. When controlling decisions are seen to be necessary, bare eco-

nomic considerations are always primary, especially when firms have hungry investors looking over their shoulders at a cupboard nearly empty of products. It is perhaps a mark of the lack of joint internal decision making and easy cooperation between scientists and managers in these firms that they can engineer such a drastic staff reduction and execute it as a complete surprise to most of the affected parties.

28. I was employed as a full-time research assistant in laboratories at Yale University (Biology) from 1970 to 1972, where I worked in an enzymology laboratory for a biochemist. I then worked briefly in the UC-Berkeley Biochemistry Department before taking a permanent laboratory position (1972–1974) in the Biophysics Group (Physiology-Anatomy) doing membrane permeability studies on erythrocytes. I left this work to enter graduate school in anthropology at Stanford University in 1975.

29. Interestingly, this first day's remarks may be the only time during the course when a "history" of the field will be addressed. Materials are usually presented as in the immediate present. See also Chapter 2.

30. The pedagogic distinction between "business" schools and "management" schools may also be reflected in the styles of comportment and interaction that differentiate each group. If we consider two polar types (when, in fact, many schools fall in between), management schools are those that largely preserve the kinds of professional divisions common to academia. Students take courses in departments that present their materials in a similar fashion to the styles of presentation of university courses and lectures, with perhaps a bit more "forced" interaction. Business schools, on the other hand, tend to have more *ad hoc* arrangements of faculty, and they teach by the "case" method, best known at Harvard. Stanford leans toward the case method in pedagogy, but preserves many traditional academic distinctions among faculty (and provides such courses for students). The Sloan School of Management at MIT leans toward the academic management pole. Many of my comments on the ambience of classrooms are more apt for the case-method environment. I have encountered few Sloan School graduates among biotechnology firm managers; but this may soon change, as PhDs in biology

began, after 1983, to come to the Sloan School for an MBA or one year Management of Technology (MoT) certificate, with the intent of looking for positions in biotech companies. Scientists from MIT, however, are legion in these companies, especially in the Cambridge/Boston area cohort.

31. This attention to record keeping in pedagogy of nascent scientists is akin to that described by Traweek in her PhD thesis on the particle physics community (1982b); see especially chap. 3, "Moving in from the Margins: The Stages of an American Career in Particle Physics." Such attention to detail and the inscription of action in records is crucial to the ongoing work of university research and industrial laboratories, and it is a quality sought in all workers from the technician level up. A recommendation that stresses these qualities is a valuable asset for advancement in the social and employment networks of academic laboratories. Commercial biotechnology labs are even more concerned with this quality, because they often use patents as a form of protection for their market niches. Establishing precedence in time for the development of a substance or process requires a careful documentation of both the process of work and the details of evidence. Inscription, in these circumstances, includes not just notebook recording (often on bound, carbon-duplicate pages), but dated, signed affirmations by laboratory heads that the inscriptions or data have been noticed and acknowledged. They are now ready for use as evidentiary "facts" in an arena of legal argument over the precedence of discovery or invention and the economic rights and protection such precedence can confer.

32. The constructivist approach to laboratory work argues otherwise, that scientific "facts" and "knowledge" are socially constructed in everyday practice. See, for instance, the citations above to Latour and Woolgar (1979) and Knorr-Cetina (1981), who examine these issues in detailed ethnographies of laboratory practice. Law and Williams (1982) also comment on the merging of economic, political, and "technical" scientific considerations in the literary and social process of constructing the text of a jointly authored article in polymer biochemistry.

33. "Decision" is from the latin root *(de) + caedere*, literally, "to cut (off)."

REFERENCES

Abir-Am, Pnina. 1982. "The Discourse of Physical Power and Biological Knowledge in the 1930's: A Reappraisal of the Rockefeller Foundation's 'Policy' in Molecular Biology." *Social Studies of Science* 12, no. 3.

Allen, Thomas J. 1966. "Studies of the Problem-Solving Process in Engineering Design." *IEEE Transactions on Engineering Management* EM-13, no. 2 (June).

———. 1977. *Managing the Flow of Technology.* Cambridge, Mass.: MIT Press.

Bailyn, Lotte. 1984. "Autonomy in the Industrial R & D Lab." Sloan School of Management, MIT Working Paper no. 1592–84.

Bateson, Gregory. 1972. "Double Bind, 1969." In Bateson, *Steps to an Ecology of Mind.* New York: Ballantine Books.

Bulkeley, William M. 1984. "Biogen's Walter Gilbert Resigns as Chief; Mark Skaletsky Named Acting Successor." *Wall Street Journal,* Dec. 18.

Cherfas, Jeremy. 1982. *Man-Made Life: An Overview of the Science, Technology, and Commerce of Genetic Engineering.* New York: Pantheon Books.

Cohen, Peter. 1973. *The Gospel According to Harvard Business School.* New York: Penguin Books.

Crick, Francis H. D. 1958. *On Protein Synthesis. Symmposium of the Society of Experimental Biology* vol. 12. Cambridge, Eng.: Cambridge University Press.

———. 1970. "Central Dogma of Molecular Biology." *Nature* 227.

Dev, Sukhendu B. n.d. "Physicists' Migration to Biology and Its Impact." Forthcoming in *Interdisciplinary Science Reviews.*

Dubinskas, Frank A. 1984. Splicing the Participant-Observer: Power and Field-Work in Genetic Engineering Firms. Paper presented at American Anthropological Association annual meeting, Nov. 17, Denver.

Dubinskas, Frank A., and Sharon Traweek. 1984. "Closer to the Ground: A Reinterpretation of Walbiri Iconography." *Man* (NS) 19.

Dyer, Gibb. 1982. "Culture in Organizations: A Case Study." Sloan School of Management, MIT, Working Paper no. 1279-82.

Fox, Jeffrey L. 1984. "Biogen Cuts 13 Percent Including Scientists." *Science* 226 (Dec. 7).

Friar, John, and Mel Horwitch. 1984. "The Current Transformation of Technology Strategy: The Attempt to Create Multiple Avenues for Innovation Within the Large Corporation." Sloan School of Management, MIT, Working Paper 1618-84.

Fuerst, John A. 1982. "The Role of Reductionism in the Development of Molecular Biology: Peripheral or Central?" *Social Studies of Science* 12, no. 2.

Fujimura, Joan. 1987. "Do-able Problems in Cancer Research." *Social Studies of Science* 17, no. 2.

"GEN Guide to Biotechnology Companies." 1983. *Genetic Engineering News* 3, no. 6.

Gilbert, G. Nigel, and Michael Mulkay. 1982. "Warranting Scientific Belief." *Social Studies of Science* 12, no. 3.

Gonick, Larry, and Mark Wheelis. 1983. *The Cartoon Guide to Genetics.* New York: Barnes and Noble.

Hochhauser, Steven J. 1983. "Bringing Biotechnology to Market." *High Technology* 3, no. 2.

Horwitch, Mel, and Kiyonori Sakakibara. 1983. "The Changing Strategy—Technology Relationships in Technology—Based Industries: A Comparison of the United States and Japan." Paper presented at Strategic Management Society annual conference, 27–29, Paris.

Judson, Horace Freeland. 1979. *The Eighth Day of Creation: The Makers of the Revolution in Biology.* New York: Simon and Schuster.

Keller, Evelyn Fox. 1983. *A Feeling for the Organism: The Life and Work of Barbara McClintock.* New York: W. H. Freeman.

———. 1984. "The Heritage of Physics in the Transformation of Biology." Paper presented at Modern Biology and the New Materialism symposium, Dec. 11, Boston Colloquium for the Philosophy of Science, Boston University.

Knorr-Cetina, Karin. 1981. *The Manufacture of Knowledge: An Essay on the Constructivist and Contextual Nature of Science.* Oxford, U.K.: Pergamon Press.

Knorr-Cetina, Karin, and Michael Mulkay, eds. 1983. *Science Observed: Perspectives on the Social Study of Science.* London: Sage Press.

Krimsky, Sheldon. 1982. *Genetic Alchemy: The Social History of the Recombinant DNA Controversy.* Cambridge, Mass.: MIT Press.

Kuhn, Thomas S. 1962. *The Structure of Scientific Revolutions.* Chicago: University of Chicago Press.

Latour, Bruno, and Steve Woolgar. 1979. *Laboratory Life: The Social Construction of Scientific Facts.* Beverly Hills, Calif.: Sage Publications.

Law, John, and R. J. Williams. 1982. "Putting Facts Together: A Study of Scientific Persuasion." *Social Studies of Science* 12, no. 4.

Milstein, Cesar. 1980. "Monoclonal Antibodies." *Scientific American,* Oct.

Mintzberg, Henry. 1973. *The Nature of Managerial Work.* New York: Harper and Row.

Old, R. W., and S. B. Primrose. 1980. *Principles of Genetic Manipulation.* Berkeley: University of California Press.

OTA Report. 1984. *Commercial Biotechnology: An International Analysis.* OTA-BA-218. Washington, D.C.: U.S. Congress, Office of Technology Assessment.

Panem, Sandra. 1984. *the Interferon Crusade.* Washington, D.C.: Brookings Institution.

Pinch, Trevor J., and Weibe E. Bijker. 1984. "The Social Construction of Facts and Artifacts: Or How the Sociology of Science and the Sociology of Technology Might Benefit Each Other." *Social Studies of Science* 14, no. 3.

Rosenfeld, Israel, Edward Ziff, and Borin Van Loon. 1983. *DNA for Beginners.* London: Writers and Readers Publishing Cooperative.

Schäfer, Wolf, ed. 1983. *Finalization in Science: The Social Orientation of Scientific Progress.* Dordrecht, Holland: D. Reidel.

Traweek, Sharon. 1982. "Gossip in Science." Paper presented at American Anthropological Association annual meeting, Dec. 6, Washington, D.C.

———. 1982b. "Uptime, Downtime, Spacetime, and Power: An Ethnography of the Particle Physics Community in Japan and the United States" Ph.D. diss., University of California–Santa Cruz.

Watson, James D. 1968. *The Double Helix: A Personal Account of the Discovery of the Structure of DNA.* New York: Atheneum.

Watson, James D., and Francis H. D. Crick. 1953. "Molecular Structure of Nucleic Acid: A Structure for Deoxyribos Nucleic Acid." *Nature* 171.

Watson, James D., and John Tooze. 1981. *The DNA Story: A Documentary History of Gene Cloning.* San Francisco: W. H. Freeman.

Wolfe, Tom. 1979. *The Right Stuff.* New York: Bantam Books.

Woolgar, Steve. 1982. "Laboratory Studies: A Comment on the State of the Art." *Social Studies of Science* 12, no. 4.

———. 1983. "Irony in the Social Study of Science." In Karin Knorr-Cetina and Michael Mulkay, eds., *Science Observed: Perspectives on the Social Study of Science.* Beverly Hills, Calif.: Sage Publications.

———. n.d. "Time and Records in Research Interaction: Some Ways of Making Out What Is Happening in Experimental Science." Unpublished MS.